Warfare in Late Byzantium, 1204–1453

History of Warfare

VOLUME 67

Warfare in Late Byzantium, 1204–1453

By

Savvas Kyriakidis

BRILL

LEIDEN • BOSTON
2011

On the cover: A View of the walls of Nicaea (Iznik).
©Photo by Savvas Kyriakidis

This book is printed on acid-free paper.

Library of Congress Cataloging-in-Publication Data

Kyriakidis, Savvas.
Warfare in late Byzantium, 1204–1453 / by Savvas Kyriakidis.
p. cm. — (History of warfare, ISSN 1385-7827 ; v. 67)
Includes bibliographical references and index.
ISBN 978-90-04-20666-3 (hardback : alk. paper)
1. Byzantine Empire—History, Military—1081–1453. 2. Military art and science—Byzantine Empire. 3. Aristocracy (Political science)—Byzantine Empire. I. Title. II. Series.

DF543.K97 2011
355.020956'0902—dc22

2011010572

ISSN 1385-7827
ISBN 978 90 04 20666 3

To my mother,
who is bravely fighting her own battle with cancer.

CONTENTS

Acknowledgments ix
Note on Transliteration xi
Abbreviations xiii
Maps xvii

Introduction 1

Chapter One Warfare and Imperial Propaganda 13
The inevitability of warfare 13
The militaristic ideology of Nicaea 16
Changes in the military profile of the emperor after the recovery of Constantinople 20
Andronikos II and the ideological importance of defensive wars against the Turks 25
The military profile of the emperor under Andronikos III and John VI Kantakouzenos 30
Fighting against impious barbarians and natural enemies 36

Chapter Two Commanders 45
The military character of the late Byzantine aristocracy 45
The role of Western European military ideals and practices 51
The effectiveness of Byzantine generals on the battlefield 60
The ideal leadership 69

Chapter Three Soldiers 75
Pronoia holders 75
The *Allagia* 82
The Nicaean frontier troops, the *Proselantai* and the *Thelematarioi* 93
The case of the *despotate* of the Morea 96

Chapter Four Mercenaries 101
The use of mercenaries 101
Categories of mercenaries 104
Byzantine attitudes towards mercenaries 118

Chapter Five Campaigns 137

Chapter Six Fortifications and Sieges 157
The role of fortifications 157
Siege warfare: The evidence of the written sources 170
The impact of gunpowder on Byzantine warfare 187

Chapter Seven Tactics 197
Late Byzantine cavalry and field tactics against Western Europeans 197
Battlefield Tactics against the Turks 203
Infantry 216

Conclusion 221

Glossary 229

Bibliography
Primary Sources 231
Secondary Sources 234

Index 245

ACKNOWLEDGMENTS

This project began as a doctoral dissertation submitted in 2007 to the Centre of Byzantine, Ottoman and Modern Greek Studies at the University of Birmingham. I am most indebted to my mentors at Birmingham, Ruth Macrides and John Haldon who introduced me to the issues and sources of Byzantine history, directed my doctoral dissertation with great care, patiently watched over the progression of various drafts and offered me a great many useful comments, as well as constructive criticism. Without their encouragement and continued interest, this project might have never been completed. I am also grateful to Dimiter Angelov for his continuous support and invaluable assistance in understanding some of the most difficult to interpret Byzantine authors. As a student of the Centre of Byzantine, Ottoman and Modern Greek Studies, I have benefited from my conversations with Anthony Bryer, the Centre's co-founder. I would also like to thank Frank Trombley for his useful suggestions and support.

This project could not have been completed without the generous postdoctoral fellowships I have been rewarded by the Research Center for Anatolian Civilizations at Koç University and the Program in Hellenic Studies at Princeton University. Both institutions provided the ideal academic environment and resources for the extensive revision of my doctoral dissertation and its transformation into a book. I owe a great deal to scholars I met in Istanbul and Princeton. To Eric Ivison, who taught me how to approach and evaluate archaeological evidence, I express my very special thanks for his encouragement, enthusiasm and genuine interest in my work. I am especially grateful to Slobodan Ćurčić, who willingly put his expertise on medieval fortifications at my disposal. I have always benefited from my long conversations with Mariya Kiprovska, Grigor Boykov and Heath Lowry who introduced me to Ottoman history and civilisation. Furthermore, I wish to thank the directors of the Research Center for Anatolian Civilizations and the program in Hellenic Studies, Scott Redford, Alessandra Ricci and Dimitri Gondicas for their useful advice, support, genuine interest in my work and for providing excellent conditions and a creative atmosphere for the successful completion of my project. I would like to thank Eve Richards for reading and editing a significant part of this work and Anastasios Tsiftsis for helping me to draw the maps.

NOTE ON TRANSLITERATION

I have employed a Greek transliteration of Byzantine names and technical terms. I also adhered to the practice of transcribing Byzantine names and not latinising them: thus Palaiologos, not Palaeologus; Kantakouzenos, not Cantacuzenus. However, some common first names have been rendered in their modern English forms: thus John, not Ioannes and Constantine, not Konstantinos. For well-known place names I have preferred the use of conventional modern English spelling: thus Athens, not Athina and Thessalonica, not Thessalonike and Gallipoli, not Gelibolu. For proper names and technical vocabulary pertaining to the Ottoman and non-Ottoman Turks, modern Turkish orthography has been used.

ABBREVIATIONS

AG	*Anecdota graeca e codicibus regiis.* 5 vols. J. Boissonade (ed.), (Paris, 1829–1833, repr. Hildesheim, 1962)
AHR	*American Historical Review*
Akropolites	*Georgii Acropolitae Opera*, 2 vols. A. Heisenberg (ed.), (Leipzig, 1903, repr. Stuttgart, 1978)
B	*Byzantion*
BF	*Byzantinische Forschungen*
BZ	*Byzantinische Zeitschrift*
BMGS	*Byzantine and Modern Greek Studies*
BSl	*Byzantinoslavica*
Chalkokondyles	*Laonici Chalcocandylae Historiarum demonstrationes*, 2 vols. E. Darkó (ed.), (Budapest, 1922–1927)
Choniates, *Historia*	*Nicetae Choniatae Historia*, J.L. van Dieten (ed.), (Berlin, 1975)
Choniates, *Orationes*	*Nicetae Choniatae orationes et epistulae*, J.L. van Dieten (ed.), (Berlin, 1973)
Chronicle of the Morea	*Chronicle of the Morea*, J. Schmitt (ed.), (London, 1904)
DOP	*Dumbarton Oaks Papers*
Doukas	*Istoria Turco-Bizantina (1341–1462),* V. Grecu (ed.), (Bucharest, 1958)
EI[2]	*The Encyclopedia of Islam*, 2nd edition
FM	*Fontes Minores*
GRBS	*Greek Roman and Byzantine Studies*
Gregoras	*Nicephori Gregorae Byzantina Historia,* 3 vols L. Schopen (ed.), (Bonn, 1829–1855)
Gregoras, *Logos*	*Nikephorou Gregora logos prosphonematikos eis ton vasilea*, in A. Westermann (ed.), *Excerptorum ex Bibliothicae Pauline Lipsiensis libris manuscriptis pars prima* (Leipzig, 1864)
Holobolos, *Orationes*	*Manueli Holoboli orationes*, M. Treu (ed.), (Potsdam, 1906–1907)
JÖB	*Jahrbuch der österreichischen Byzantinistik*
JGR	*Jus Graecoromanum*, 8 vols. P. and I. Zepos (eds.), (Athens, 1931, repr. Aalen, 1962)
JMH	*Journal of Medieval History*
Kananos	*Johannis Canani. De Constantinopolis obsidione*, E. Pinto (ed.), (Napoli, 1990)

Kantakouzenos	*Ioannis Cantacuzeni eximperatoris historiarum libri IV*, 3 vols. L. Schopen-B. Niehbur (eds.), (Bonn, 1828–1832)
Kydones, *Correspondance*	*Démétrius Cydonès correspondance*, 2 vols. R.-J. Loenertz (ed.), (Vatican, 1956, 1960)
Lampenos, *Encomium*	I. Polemis, *Ο λόγιος Νικόλαος Λαμπηνός και το εγκώμιον αυτού εις τον Ανδρόνικον Β΄ Παλαιολόγον* [The scholar Nikolaos Lampenos and his Encomium to Andronikos II] (Athens, 1992)
Manuel II, *Funeral Oration*	*Manuel II Palaiologos. Funeral Oration on his Brother Theodore*, J. Chrysosteomides (ed.), (Thessalonica, 1985)
Manuel II, *Letters*	*The Letters of Manuel II Palaeologus*, G.T. Dennis (ed.), (Washington, 1977)
Matthew of Ephesos, *Briefe*	*Die Briefe des Matthaios von Ephesos im Codex Vindobonensis Theol. Gr. 174*, D. Reinsch (ed.), (Berlin, 1974)
Metochites, *Miscellanea*	*Theodori Metochitae miscellanea philosophica et historica*, G. Müller-T. Kiessling (eds.), (Leipzig, 1821, repr. Amsterdam, 1966)
Metochites, *Orations*	*Οι δύο Βασιλικοί Λόγοι* [The Two Imperial Orations], I. Polemis (ed.), (Athens, 2007)
MHR	*Mediterranean Historical Review*
MM	*Acta et diplomata graeca medii aevi*, 6 vols. F. Miklosich–J. Müller (eds.), (Vienna, 1860–1890)
MP	*Monumenta Peloponnesiaca. Documents for the History of the Peloponnese in the 14th and 15th Centuries*, J. Chrysostomides (ed.), (Camberley, 1995)
Muntaner	Muntaner Ramon, *Crònica*, 2 vols. V.J. Escartí (ed.), (Valencia, 1999)
OCP	*Orientalia Christiana Periodica*
Pachymeres	*Georges Pachymérès. Relations historiques*, 5 vols. A. Failler (ed.), (Paris, 1984–2000)
PG	J.-P. Migne, *Patrologiae Cursus completus, series Graeca*, 161 vols. (Paris, 1857–1866)
Philes	*Manuelis Philae carmina*, 2 vols. E. Miller (ed.), (Paris, 1855–1857)
PL	J.-P. Migne, *Patrologiae cursus completes, series Latina*, 221 vols. (Paris, 1844–1864)
PLP	Prosopographisches *Lexikon der Palaiologenzeit*, 12 vols. E. Trapp (ed.), (Vienna, 1976–1996)
PP	*Παλαιολόγεια και Πελοποννησιακά* [*Palaiologeia kai Peloponnesiaka*], 4 vols. S. Lampros (ed.), (Athens, 1912–1930)
Pseudo-Kodinos	*Traité des offices*, J. Verpeaux (ed.), (Paris, 1966)
REB	*Revue des Études byzantines*

REI	*Revue des Études Islamiques*
Sanudo	*Marin Sanudo Torsello. Istoria di Romania*, E. Papadopoulou (ed.), (Athens, 2000)
Sphrantzes	*Giorgio Sfranze Cronaca*, R. Maisano (ed.), (Rome, 1990)
Strategikon	*Das Strategikon des Maurikios*, G.T. Dennis (ed.), (Vienna, 1981)
SüdostF	*Südost Forschungen*
Taktika	*The Taktika of Leo VI*, G.T. Dennis (ed.), (Washington, 2010)
Theodore II, *Epistulae*	*Theodori Ducae Laskaris Epistulae CCXVII*, N. Festa (ed.), (Florence, 1898)
Thiriet, *Assemblées*	F. Thiriet, *Délibérations des assemblées vénitiennes concernant la Romanie*, 2 vols. (Paris, 1966–1971)
Thiriet, *Régestes*	*Régestes des délibérations du Sénat de Venise concernant la Romanie*, 3 vols. (Paris, 1958–1961)
TM	*Travaux et Mémoires*
VyzSym	*Byzantina Symmeikta*
ZRVI	*Zbornik Radova Vizantološkog Instituta*

MAPS

Map 1: Asia Minor xviii
Map 2: The Balkans xix

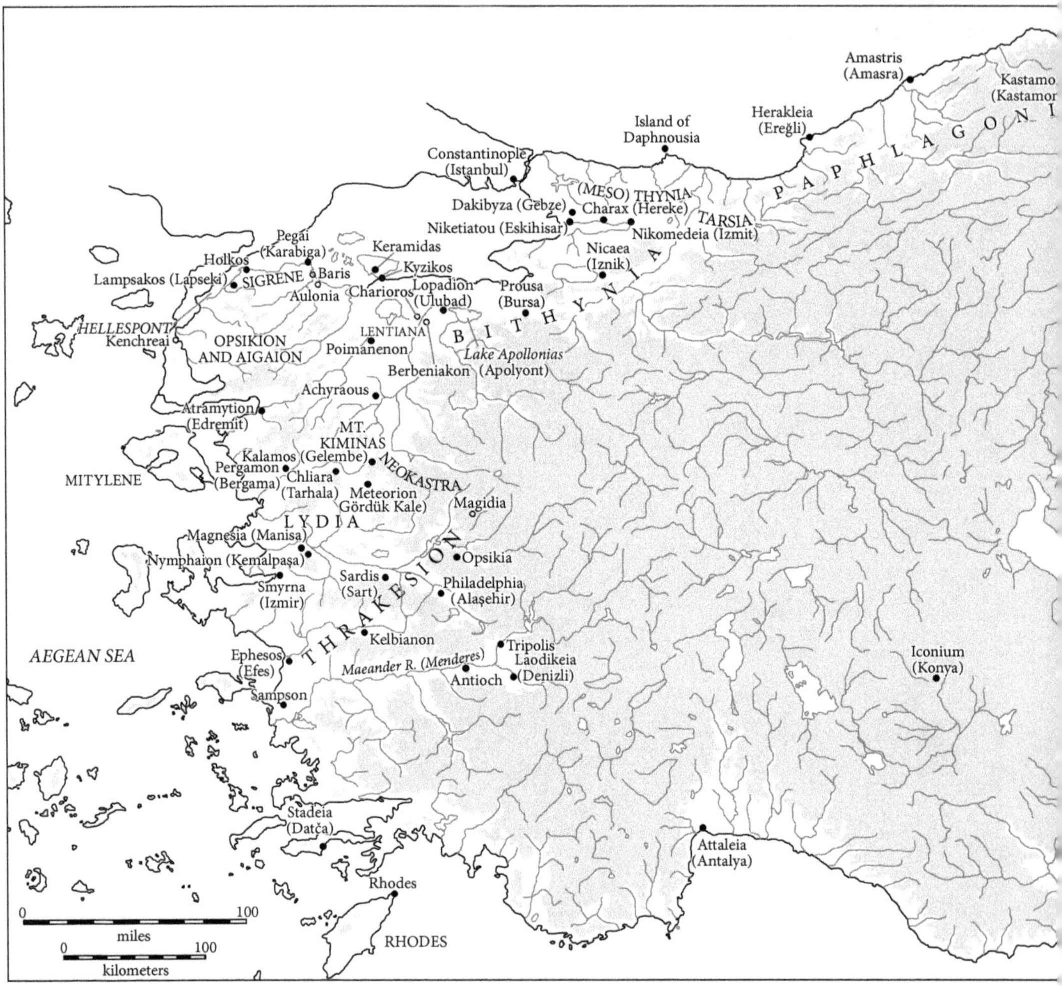

Map 1: Asia Minor

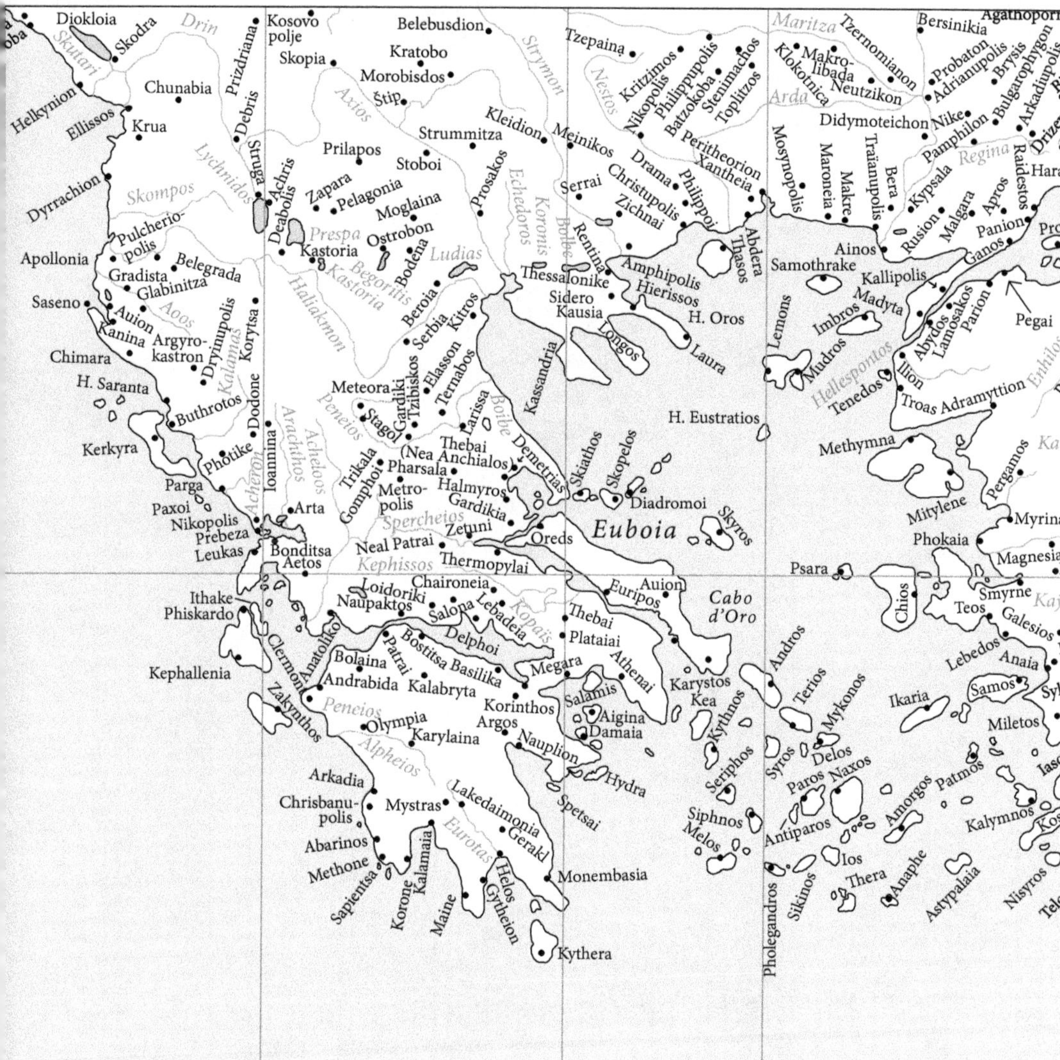

Map 2: The Balkans

INTRODUCTION

The capture of Constantinople by the armies of the Fourth Crusade in 1204 fragmented the Byzantine empire. Territories which did not submit to the Crusaders fell into the hands of Byzantine magnates who became rulers of numerous small political entities. The most important of these newly founded states, which each claimed to be the successor of the destroyed Byzantine empire, were the empires of Trebizond and Nicaea in Asia Minor and the principality of Epiros in the Balkans.[1] The so-called empire of Nicaea, which was established as a viable state by Theodore I Laskaris (1204–1221), was the most successful of these. Laskaris averted the threat of a combined attack from the Latin empire of Constantinople and the Seljuks of Rum and overcame various local lords who, in the wake of the collapse of Byzantium, had established their own independent authorities in Asia Minor. His successors, John III Vatatzes (1221–1254) and Theodore II Laskaris (1254–1258), conquered large territories in the Balkans. They forced the rulers of Epiros to abandon their claim to the imperial title and reduced the military strength of the Latin empire of Constantinople. Under John III and Theodore II, Nicaea prevailed as the legitimate successor to the Byzantine empire. In 1261, the Nicaean army captured Constantinople and Michael VIII Palaiologos (1259–1282), who had seized the throne from the Laskarids, restored the Byzantine empire. Yet, despite the conquests of the Laskarids and the recovery of Constantinople by Michael VIII, much territory which had belonged to the Byzantine empire before the Fourth Crusade remained beyond imperial control. Many Aegean islands, most of the Peloponnese, a large part of mainland Greece and the Ionian islands were ruled by Western European lords. Although the rulers of Epiros and Thessaly ceased to call themselves emperors, their states remained effectively independent territories.

[1] See N. Oikonomides, "La décomposition de l' empire Byzantin à la veille de 1204 et les origines de l' Empire de Nicée: À propos de la *Partitio Romaniae*," in Actes du *XV*[e] *Congrès des études byzantines. Rapports et co-rapports*, I, (Athens, 1976), 1–28, reprinted in N. Oikonomides, *Byzantium from the Ninth Century to the Fourth Crusade. Studies, Texts, Monuments* (Aldershot, 1992), Study XX.

The reign of Michael VIII's successor, Andronikos II Palaiologos (1282–1328) witnessed the loss of Asia Minor to the Turcoman principalities and the increasing strength of the Serbian kingdom, which grew in Macedonia, Epiros and Albania at the expense of the Byzantines. Andronikos II was overthrown by his grandson Andronikos III (1328–1341) as a result of a civil war waged intermittently from 1321 to 1328. Andronikos III's death in 1341 was followed by a catastrophic civil war between his close associate, John Kantakouzenos (John VI, 1347–1354), and the regency of the legitimate successor to the throne, John V Palaiologos (1341–1391). In 1352, John V rose against Kantakouzenos, who was forced to abdicate two years later. During the conflicts between the clans of the Kantakouzenoi and the Palaiologoi, the Serbian ruler, Stefan Dušan (1331–1355) exploited the weakness of the Byzantines and captured almost all the Byzantine possessions in Albania, Epiros, Thessaly and most of Macedonia. Consequently, the Byzantine state was reduced in Thrace and Constantinople to a small district around Thessalonica, the despotate of Morea in the Peloponnese and a small number of northern Aegean islands. Each of these territories constituted a de facto semi-independent principality ruled by members of the extended imperial family. Dušan's son and successor, Stefan Uroš (1355–1371), was finally unable to maintain the unity of his father's dominions. As a result, the Serbian empire disintegrated and a multitude of small centres of power was established in the Balkans. The emperor in Constantinople, together with the despots of the Morea and Thessalonica, belonged to this extremely fragmented world of powerful and ambitious men who, in their small principalities, were short of authority. These states were constantly at war with each other over ill-defined frontiers and territorial differences, the limited available resources and local pre-eminence. They lacked any elaborate administrative infrastructure and their unity and survival relied on the personality and wealth of their rulers, in particular on their ability to impose their authority over local magnates and recruit effective military forces to defend their lands. The fate of many of these states was identified with that of their rulers. This fragmented political situation when they had already established themselves as the dominant power in Asia Minor, helped the Ottomans advance into the Balkans.

Therefore, the period 1204–1453 was characterised by political fragmentation, endless wars between small political entities and continuous crises. It is in this respect that the study of Byzantine military thought and attitudes towards war is important. They reflect the continuous struggle of small states to survive and defend themselves against large numbers of hostile neighbours, some of whom had or gradually acquired the resources to develop military forces far superior to those of Byzantium. The study of late Byzantine military institutions and military thought reflects also the attitudes of a state which, aware of its limited resources, wanted to avoid military conflict, but in response to the conditions of the time considered warfare unavoidable and the maintenance of a strong army led by a soldier emperor essential for survival.

Regarding the arrangement of the material in the present investigation, the first chapter examines how late Byzantine rulers used imperial propaganda to justify the wars that they fought and to promote their military image, their military policies and their priorities. Moreover, the discussion of the role of the imperial office in the military affairs of the empire provides an insight into the relationship between the throne and the aristocracy, the members of which monopolised military command. The second chapter deals with the question of military leadership in late Byzantium. It investigates the military character and ethos of the higher aristocracy and compares them to concurrent developments in Western Europe. It also examines the effectiveness and role of late Byzantine generals on the battlefield. The third and fourth chapters deal with the categorisation of soldiers into various groups and examine their effectiveness on the battlefield. They also discuss the views expressed in the sources concerning the various types of soldier, in particular the mercenaries, who were recruited from outside the empire and dominated the military conflicts of the period under discussion. The fifth chapter examines the organisation and supply of the military campaigns conducted by the late Byzantine armies. The sixth chapter discusses the role of fortifications in the defence of the late Byzantine empire, and the nature of siege warfare from the thirteenth through fifteenth centuries. The seventh chapter investigates the tactical effectiveness of the late Byzantine armies. It explores the role and use of infantry and cavalry, how the late Byzantines adjusted their military tactics to take into account the nature of their enemies and the Byzantine perceptions about the nature and strength of their enemies.

A question which is treated in each part of the present work is the Western European influences on Byzantine attitudes towards warfare, military administration and the conduct of fighting. Late Byzantine military developments should be examined in the general context of contemporary developments in Western Europe, the Balkans and in the East. Western European military practices had influenced Byzantium since the Komnenian period. It might be expected that these influences would be more apparent in the late Byzantine period. The development of trade increased the contacts between Byzantium and the West, while the Crusades and the establishment of Latin and Turcoman principalities in old Byzantine territories brought foreign military cultures closer to Byzantium.

The most important constraint for the study of Byzantine military history after 1204 is the absence of military manuals, which are the most valuable sources for the study of warfare from the sixth to the eleventh centuries. The present investigation makes use of the military treatises which were compiled in earlier periods of Byzantine history, in order to examine how far the principles and instructions prescribed by their authors were followed in the later centuries. The only contemporary treatises used for the aims of the present work are those of the marquis of Montferrat, Theodore Palaiologos (1291–1338), and of Pseudo-Kodinos.[2] However, neither of these can be seen as a strictly military treatise. Theodore was Andronikos II's second son by Yolande of Montferrat. In 1306, he was sent to rule Montferrat and in 1326 compiled a treatise on the art of governing and military strategy. Since he lived almost all of his adult life in the West, where he acquired his entire first-hand military experience, his treatise is heavily influenced by the context of warfare in fourteenth-century Italy.[3] Nevertheless, Theodore's work is useful for comparisons between Byzantine and Italian practices and attitudes towards warfare. The treatise of Pseudo-Kodinos, a work on court hierarchy and court ceremony, was compiled in the mid-fourteenth century. Its anonymous author provides useful information concerning the function of military offices, the role of imperial guard units and the organisation of military campaigns.

The absence of military treatises means that most of our knowledge of late Byzantine military history derives from the narrative sources. The most important of these is the *History* of John Kantakouzenos

---

[2] For Theodore Palaiologos see *PLP*, 21465. For the treatise of Pseudo-Kodinos as a source of late Byzantine history see R. Macrides, "The Reason is not Known. Remembering and Recording the Past. Pseudo-Kodinos as a Historian," in P. Odorico – P. Agapitos – M. Hinterberger (eds.), *L'écriture de la mémoire. La littérarité de la historiographie* (Paris, 2006), 317–330.

[3] See A. Settia, *De re militari: pratica e teoria nella guerra medievale* (Rome, 2008), 113–152.

(ca. 1295–1383). His work, which covers the period 1320–1362 and was compiled in the 1360s, is a detailed narrative of the political and military events of the period and the main source for the civil wars and conflicts between the cliques of the social elite which dominated the first half of the fourteenth century.[4] Having been the supreme commander of the Byzantine army and an emperor for more than three decades, Kantakouzenos provides information on every area that this work touches, ranging from the justification of warfare to Byzantine attitudes towards foreign mercenaries, and from battlefield tactics to the social distinctions within military forces. However, most of the information provided by Kantakouzenos on these questions is incidental. His aim was not to describe the Byzantine administrative system, but to explain his view of the events which ensued from the 1320s until the 1350s, in which he played a leading role. Justifying his support for Andronikos III in the civil war of 1321–1328, he presents him as a great general, while his image of Andronikos II is very negative. Kantakouzenos was present at almost every military operation he describes. However, his tendency to overemphasise his personal influence on the outcome of military conflicts prevents us from reaching safe conclusions on the nature of the military conflicts in his account.

After Kantakouzenos, the most useful sources for the study of late Byzantine history, of the thirteenth century in particular, are George Akropolites (1217–1282) and George Pachymeres (1242–ca. 1310). Akropolites' work is the main source for the history of the so-called empire of Nicaea. Warfare dominates his account.[5] However, it was only after 1246, when he started accompanying the emperors on their campaigns in the west, that he began to provide circumstantial details concerning the composition, strength and effectiveness of the Nicaean army. Akropolites constantly tries to dismiss Theodore II's military skills and expresses rather negative views about the generals whom Theodore II elevated to high office, although he himself was one of them. He was the only one of the new men thus promoted

---

[4] On the dating of the *History* of Kantakouzenos see H. Hunger, *Die hochsprachliche profane Literatur der Byzantiner*, 2 vols.(Munich, 1978), I, 469; D. Nicol, *The Byzantine Family of Kantakouzenos (Cantacuzenus) ca. 1100–1460* (Washington, 1986), 100.

[5] R. Macrides, "The Thirteenth Century in Byzantine Historical Writing," in C. Dendrinos *et al.* (eds.), *PORPHYROGENITA. Essays on the History and Literature of Byzantium and the Latin East in Honour of Julian Chrysostomides* (London, 2002), 65.

who survived Michael VIII's usurpation and therefore it was too dangerous for him to associate himself with the policies of Theodore II.[6] Consequently, Akropolites' description of Theodore II's military policies and campaigns should be treated cautiously. The account in the *Synopsis Chronike*, which is a chronicle dealing with the events from the Creation of the World to 1261 and commonly attributed to the metropolitan of Kyzikos, Theodore Skoutariotes, complements the *History* of Akropolites. The author's version of the history of Nicaea relies on Akropolites, without, however, sharing the latter's negative description of the Laskarids. He adds details concerning the military operations of Theodore II which Akropolites omits and presents a more positive view of his reign.[7]

The work of George Pachymeres covers the period 1250s–1307. Although a cleric in the patriarchical hierarchy, Pachymeres possessed substantial secular knowledge and put forth economic reasons as an explanation for the military developments of the period, as well as the attitudes and activities of the soldiers. Moreover, his account complements and contradicts the *History* of Akropolites.[8] Pachymeres provides useful information regarding the evolution and strength of the army and also of the impact of internal political and social conflicts on the effectiveness of the Byzantine military forces. What makes Pachymeres particularly interesting is that he provides a reasoned explanation for the disintegration of Byzantine control over Asia Minor and the effect which this development had on the army. It should be noted, however, that his critical attitude to the fiscal policies of Michael VIII and Andronikos II seems to have made him idealise the Nicaean past and he uses Theodore II as a model ruler to contrast with the first two Palaiologoi. The account of Nikephoros Gregoras which covers the period 1204–1359 complements the information provided by Pachymeres and Kantakouzenos. Unlike Kantakouzenos, Gregoras was a supporter of Andronikos II. He objected to many of the policies of Andronikos III, whom he repeatedly portrays as a poor and inexperienced general. Furthermore, Gregoras provides us with

---

6 For a detailed discussion see R. Macrides, *George Akropolites. The History* (Oxford, 2007), 19–27, 58–59.

7 Macrides, "Thirteenth Century," 69; idem, *The History*, 65–71; For the possibility that Skoutariotes was not the author of the *Syvopsis Chronike* see K. Zafeiris, "The Issue of the Authorship of the *Synopsis Chronike* and Theodore Skoutariotes," in *REB* 69 (2011), forthcoming.

8 Macrides, *The History*, 75.

important information about the battlefield tactics employed by the Byzantines and their enemies and describes in detail the difficulties of the Byzantine armies in dealing effectively with the fighting techniques of their opponents.

The fifteenth-century chronicles were written after the fall of Constantinople in 1453. The work of George Sphrantzes covers the period 1401–1477 and is compiled in the form of annalistic memoirs. Sphrantzes was a high-ranking court official who served at the court of Manuel II (1391–1425), John VIII (1425–1448) and Constantine XI (1448–1453). He also held positions in the Morea and undertook many diplomatic missions. He fought on the side of Constantine Palaiologos in the Morea and was present at the fall of Constantinople. However, he provides very little information about the military affairs of the empire. The *Histories* of Doukas, which cover the period 1341–1462 (only after 1389 in detail), and of Laonikos Chalkokondyles which narrate the events of the period 1298–1464 focus on the rise of the Ottomans. They say very little about the Byzantine armies. Nonetheless, they are useful sources for our understanding of the development of the Ottoman army and the nature of fifteenth-century warfare.

Non-Byzantine chronicles compiled in territories which had belonged to the Byzantine empire before 1204 and which place Byzantine warfare in the wider geographical and cultural context constitute a significant source for the present study. The most important of them is the fourteenth-century *Chronicle of the Morea*, which has survived in three different versions. It makes lengthy comparisons between the battlefield tactics used by the Byzantines and the Franks and provides information on the similarities and differences between the Franks and the Byzantines with regard to conduct on the battlefield and attitudes towards warfare. The fifteenth-century anonymous *Chronicle of the Toccos* is an excellent source of material on the small-scale warfare between the small principalities in fifteenth-century Greece and Albania. Its main subject is the military deeds of the ruler of Cephalonia, Carlo Tocco, who for a short time extended his dominions into the Peloponnese. This anonymous work illustrates the advanced political and regional fragmentation of the Byzantine world during the fifteenth century.[9] The *Chronicle* of Ramon Muntaner, one

[9] P. Magdalino, "Between Romaniae: Thessaly and Epiros in the Late Middle Ages," in D. Arbel – D. Jacoby (eds.), *Latins and Greeks in the Eastern Mediterranean after 1204* (London, 1989), 90.

of the leading members of the Catalan Grand Company which was employed by the Byzantines in the first decade of the fourteenth century is very useful for the study of the role of foreign mercenaries in Byzantium and the warfare between the Byzantines and the Turks in Asia Minor. It would be impossible to fully understand the conflict between the Byzantine government and the Catalans without comparing the accounts of Pachymeres and Gregoras to the information provided by Muntaner. The fourteenth-century Chronicle entitled *Storia di Romania* was compiled by the Venetian Marino Sanudo; he spent much of his life in the possessions of Venice in the Aegean and can complement the information provided by the sources listed above.

The reliance on narrative histories to reconstruct the military history of the later Byzantine empire means that certain constraints should be admitted. Although matters of war dominate the historical accounts of the period, most late Byzantine authors had no military experience; they did not personally observe the warfare between Byzantium and its enemies. Consequently, their descriptions of battles and campaigns are frequently vague and summary. They pay little attention to such matters as tactics, terrain, armaments or logistics and they relate rather little about military organisation and war making. Most of the information they provide is of a circumstantial nature. Moreover, the fact that a knowledge of classics often took precedence over the accurate and precise description of military operations meant that the major historians of the late Byzantine period made rather little use of contemporary technical terminology. For instance, the use of the terms *strateumata, dynameis, stratopeda, stratia, tagmata* and *taxeis*, to refer indiscriminately to the various army units prevents us from fully understanding the internal organisation of the late Byzantine military forces. This inconsistency makes it difficult to study thoroughly the meaning, role and internal differences of the *allagia*, a term which appears mainly in documentary sources and in chronicles written in a language close to the contemporary spoken language. Similarly, the use of generic phrases such as, 'war engines of every kind,' prevents us from reaching safe conclusions concerning the nature of siege warfare during the period under discussion. Moreover, almost all historical accounts were compiled in Constantinople and tend to report events which took place near the capital. Therefore, they provide very little information concerning Byzantine military operations and arrangements in territories such as Epiros and the Peloponnese.

The large number of imperial orations (*vasilikoi logoi*) which were compiled in the late Byzantine period, together with funeral orations and speeches addressed either to the residents of an individual city or to the Byzantine people in general, provide the backbone for the study of the ideological aspects of warfare.[10] As was the case with all imperial prerogatives in late Byzantium, the authority of the emperor to wage war was not defined by legal scholars. Instead, rhetoric was the main means through which political messages were conveyed and political thought was expressed.[11] Byzantine orators followed rhetorical models which were established long before the thirteenth century and the rules of the genre meant that imperial orations should praise the emperor, mention what the ruler was supposed to achieve by virtue of his office and omit failures. Nonetheless, imperial panegyrics should not be treated merely as attempts by the authors to praise their emperors and patrons. As George Dennis has pointed out, unlike narrative histories, imperial orations were compiled shortly after the events which they recorded and since people in the audience had played an active role in these events, imperial orations cannot be complete fabrications.[12] The analysis of the allusions to contemporary events which can be found in the encomia lead to useful conclusions concerning the image of the emperor as warrior and the justification of the military conflicts in which the late Byzantine armies were involved. Furthermore, through panegyric late Byzantine orators found the opportunity to voice their opinions on important public issues.[13]

The letters compiled by Byzantine emperors such as Theodore II and Manuel II or important statesmen such as Demetrios Kydones (1323/24–1397/98), who was the *mesazon* of John VI, John V and Manuel II, provide scanty but useful information about the military affairs of the empire. However, the fact that the authors felt compelled

---

[10] For the imperial orations in the later Byzantine period see D. Angelov, "Byzantine Imperial Panegyric as Advice Literature," in E. Jeffreys (ed.), *Rhetoric in Byzantium* (Aldershot, 2003), 55–72; idem, *Imperial Ideology and Political Thought in Byzantium. 1204–1328* (Cambridge, 2007), 51–77; T. Kiousopoulou, *Βασιλεύς ή οικονόμος; Πολιτική Εξουσία και Ιδεολογία πριν την Άλωση* [*Emperor or Administrator? Political Power and Ideology before the Fall*] (Athens, 2007), 164–181.

[11] Angelov, *Imperial Ideology*, 17.

[12] G.T. Dennis, "Imperial Panegyric: Rhetoric and Reality," in H. Maguire (ed.), *Byzantine Court Culture from 829 to 1204* (Washington, 1997), 137.

[13] Angelov, "Imperial Panegyric," 58–72.

to follow the rules of epistolography meant that the large number of letters compiled by emperors who led military forces are of little use for the study of Byzantine warfare.[14] For instance, Theodore II avoids being specific and the actual events he refers to are veiled in metaphor. Manuel II Palaiologos makes it clear that epistolography is not the best source from which to reconstruct historical events. Being a vassal of the Ottoman ruler Bayezid I (1389–1402), Manuel II followed him in one of his expeditions in Asia Minor in the winter of 1391. In a letter he compiled during this campaign, Manuel refers to the hardships and difficulties that the army had to face due to campaigning out of season. However, he does not provide any detailed information, for, as he comments, the hardships they suffered call for a historian not a letter-writer.[15]

Documentary sources, such as the monastic archives and the registers of the patriarchate of Constantinople, could have provided the bulk of information concerning the maintenance and categorisation of late Byzantine soldiers, as well as the social differentiation in the military forces and the status of soldiers in the wider society. However, very few of them have come down to us and hence relying on this type of source to reconstruct the organisation and nature of the army could lead to over-generalisation and misleading conclusions.

Archaeological evidence provides a valuable insight into Byzantine strategy, military organisation and siege warfare. The investigation of siege warfare in the later Byzantine period is essential for understanding the effectiveness of the Byzantine armies, since sieges and the defence of strategic cities formed an important element of Byzantine war planning and strategic thought. The written sources indicate that sieges and blockades of cities and fortresses substantially outnumbered pitched battles. Therefore cross referencing between written material and archeological findings is essential in understanding the nature of sieges during the period under discussion, the technological developments in warfare and the role of fortifications in late Byzantium.

The study of late Byzantine warfare is in many respects still an underdeveloped field. This reflects the scarcity of the surviving sources and, perhaps, a general view that in late Byzantium the army was composed of various ineffective foreign and disloyal mercenaries, who were hast-

---

14 For epistolography in Byzantium see Hunger, *Literatur*, I, 199–239.
15 Manuel II, *Letters*, 59.

ily recruited whenever the need arose, and that all the operations the Byzantine armies undertook were defensive and reflected desperate responses to enemy advances. This interpretation can be seen as part of a wider tendency to examine later Byzantine history as a period of the decline of a long enduring civilisation. However, this approach reinforces the perception of an unchanged and monolithic Byzantium which was doomed to destruction, which prevents modern scholarship from placing late Byzantine society and culture, and therefore warfare, in the context of contemporary developments in late medieval Europe, in the Balkans and in the East.[16]

There is only one monograph dealing with the subject. Mark Bartusis produced in 1992 a survey and analysis of the history of the late Byzantine army. Beyond the history of the army, Bartusis' study focuses on the military organisation and administration of the Byzantine empire, on the troops of various kinds, their numbers and their financing. It is the only social and administrative study dedicated to the late Byzantine army. Aside from the above-mentioned work, modern scholarship has produced a limited number of articles dealing with aspects of military administration, such as Bartusis' discussion on the smallholding soldiers in late Byzantium, Oikonomides' contribution to the study of financing of the soldiers of the early Palaiologan period and Panagiotides' examination of the role of storehouses and markets in supplying military campaigns.[17] Thus, very little research has been done in respect of the study of war making, the examination of the military ethos and attitudes towards warfare, the effectiveness of the various groups of soldiers fighting for Byzantium, the role of mercenary soldiers, the nature and role of military leadership and the influence which other medieval cultures exerted on Byzantine military thought, organisation and practice.

[16] See the discussion of Kiousopoulou, *Βασιλεύς ή Οικονόμος*, 11–12.

[17] N. Oikonomides, "À propos des armées des premières Paléologues et des compagnies de soldats," *TM* 8 (1981), 353–371; Bartusis, "On the Problem of Smallholding Soldiers in Late Byzantium," *DOP* 44 (1990), 1–26; K. Panagiotides, "Η επιμέλεια του στρατού κατά τους ύστερους Βυζαντινούς χρόνους (1204–1453)," [The Supplying of the Army in the Late Byzantine Period] *Byzantina* 24 (2004), 289–313.

CHAPTER ONE

WARFARE AND IMPERIAL PROPAGANDA

The inevitability of warfare

The period 1204–1453 was a time of turmoil and endless wars. This context, in which war and not peace was seen as the natural state of the numerous small political entities of the Balkans and Asia Minor, led Byzantine authors to view military conflict as something unavoidable. They also conclude that in a politically fragmented world diplomacy could not bring peace. The philosopher, author and close associate of Andronikos II, Theodore Metochites led many peace negotiations and diplomatic missions in Serbia. However, in his writings he opposes any diplomatic dealings with the Turcoman principalities in Asia Minor. He considers the Turks as a nomadic people, a fluid society, lacking any form of centralised political organisation. Had the Byzantines been dealing with a centralised state, as he argues, diplomacy could have led to a solution. However, since the numerous Turcoman principalities were small political entities, diplomatic dealings with them were of no avail.[1] In another of his works, Metochites comments that no one denies that living in peace is evidence of Christian piety. It is during peace that the rule of law prevails and the state increases its incomes and wealth.[2] However, war is inevitable and fighting is the result of human greed. Therefore, to achieve peace it is necessary to be ready for war, for and those who believe that they will live permanently in peace do not succeed in doing so and they either succumb to slavery or they are forced to fight wars. Metochites concludes that all people love easy profit and to gain this they carry out attacks, in particular on those whose minds are fixed on peace. Therefore, to enjoy the benefits of peace and justice it is necessary for even the civilian

[1] Metochites, *Orations*, 384.

[2] Metochites, *Miscellanea*, 511–512. For Metochites' career see I. Ševčenko, *Études sur la polémique entre Théodore Métochite et Nicéphore Choumnos* (Brussels, 1962), 129–144; idem, "Theodore Metochites, the Chora, and the Intellectual Trends of His Time," in P.A. Underwood (ed.), *The Kariye Djami. Studies in the Art of the Kariye Djami and its Intellectual Background*, 4 vols. (Princeton, 1975), IV, 19–55.

population (*politikon*) to be ready to fight wars.[3] That military conflict was considered inevitable, while diplomacy could not ensure peace is expressed in the writings of the monk, teacher and diplomat Maximos Planoudes.[4] In the imperial oration which he compiled in 1294 for the coronation of Michael IX (1294–1320) as co-emperor of Andronikos II, Planoudes observes that the emperor should always suspect the motives of his enemies when conducting diplomatic negotiations. He adds that the emperor should possess a sizeable army, including even peasants and shepherds, a large native and mercenary cavalry force, a powerful fleet and a great military budget. Planoudes adds that the emperor should be a competent soldier, so that not only will his subjects know his virtues, but also the enemies will be astonished by his military skills.[5] Like Metochites, Planoudes' views about war and diplomacy were influenced by the conflict between Byzantium and the Turcoman principalities in Asia Minor.

The idea that the empire was surrounded by enemies and needed to be constantly ready for war and ruled by a warrior emperor is emphasised in the *Mirrors of Princes* compiled by late Byzantine authors. In 1250, the polymath clergyman and author Nikephoros Blemmydes, who was the tutor of the emperor Theodore II Laskaris, compiled his *Imperial Statue.* In this work, Blemmydes offers advice on military training and tactics, among other things. He also points out that the emperor needs to be constantly ready for war, since wars can flare up quite unexpectedly. Fighting and training in the art of war should be one of the most important activities of the emperor and his army.[6] Warfare is presented as an integral part of the ruler's life in another *Mirror of Princes* which was compiled by the monk Theognostos and was chronologically close to that of Blemmydes. Theognostos exhorts the emperor to do everything possible to avoid bloodshed. When treaties and agreements are possible, the emperor should readily undertake negotiations for concord in a spirit of charity.[7] In the first decade of the fourteenth century, the philologist and teacher from Thessalonica, Thomas Magistros compiled a work in which he states that most battles

---

[3] Metochites, *Miscellanea*, 516–520.

[4] For the career of Maximos Planoudes see *PLP*, 23308.

[5] G. Westernik, "Le basilikos de Maxime Planoude," *BSI* 27 (1966), 60, *BSI* 28 (1968), 44–45.

[6] *PG* 142, cols. 638–640.

[7] *Theognosti Thesaurus*, J. Munitiz (ed.), (Brepols, 1979), 198.

are the result of injustice. Due to injustice inflicted upon the empire, the emperor is forced to get involved in battles to defend the law. It is impossible to enjoy peace without being warlike (*philopolemos*). Furthermore, Magistros praises the people who are trained in the art of war and concludes that being surrounded by so many enemies the emperor is forced to be warlike if he wants peace.[8]

It is not surprising that in this context of continuous and inevitable warfare the imperial office was heavily militarised. This was not something new. The militarisation of the imperial office is closely associated with the emergence of the military aristocracy as the ruling elite of Byzantium in the eleventh century, when the emperors started to be praised as dedicated warriors. Until then, with the exception of Nikephoros II Phokas (963–969), no martial virtues of the imperial image were praised. Instead, the emperor was usually presented as a peace-maker.[9] Nonetheless, in late Byzantium the military virtues of emperors were emphasised far more intensively than in the past.[10] Moreover, throughout Byzantine history war was a state issue and it was the emperor's prerogative to wage it as the instigator of every aspect of policy in the Byzantine state. It was the ruler's duty to organise the defence of the empire and command its army. In the preamble of a *chrysobull* issued by Andronikos II after 1294, it is stated that it is an imperial duty to look after the military forces of the empire, to organise the military rolls and make the necessary arrangements for the provision of weapons, so the empire would be able to deploy a large military force against its enemies.[11]

Rhetoric was the means by which Byzantine rulers promoted their military profile and advertised their military priorities. The rhetorical conventions for celebrating an emperor which are outlined in a handbook compiled by the second-century author known as 'Menander Rhetor' emphasise the promotion of the military virtues of the ruler. Menander writes that the orator must describe the emperor's own battles, his engagement with the enemy, prowess and deeds of courage, as well as his armour and campaigns. In the fourteenth century Joseph

[8] *PG* 145, col. 457.

[9] A. Kazhdan – A.W. Epstein, *Change in Byzantine Culture in the Eleventh and Twelfth Centuries* (Berkeley, 1985), 102–113; P. Magalino, *The Empire of Manuel I Komnenos. 1118–1143* (Cambridge, 1993), 418–423.

[10] Angelov, *Imperial Ideology*, 83.

[11] MM, V, 264. For the dating of this document see Angelov, *Imperial Ideology*, 37.

the Philosopher also advised panegyrists to praise the excellence of the emperor's armour.[12] In addition, the examination of imperial propaganda and rhetoric shows that different emperors had different priorities which often were the opposite of their predecessors. For instance, Michael VIII had concentrated on the conquest of western territories which had belonged to the empire before 1204. Consequently, imperial propaganda emphasises the ideological importance of his wars in the west which are presented as Just Wars for the recovery of imperial territory. His successor Andronikos II followed a different policy and concentrated his military efforts on the defence of Asia Minor against the Turks. Therefore, during his reign defensive wars against the Turcoman principalities became ideological important.

The militaristic ideology of Nicaea

Under the Laskarid dynasty, the promotion of the image of the emperor as a great warrior was amplified. It is closely connected to the cultivation of an ideology of militarism and reconquest in the newly founded state of Nicaea which is an immediate ideological response to the capture of Constantinople by the armies of the Fourth Crusade in 1204.[13] This militaristic ideology relied on Old Testament ideas. The fall of Constantinople was compared to the 'Babylonian captivity' and all the Laskarid emperors of Nicaea are presented as the rulers who would lead the return of their people in Constantinople.[14] Moreover, warfare could be considered a holy activity. In an act issued between 1208 and 1210, the patriarch Michael Autoreianos granted full remission of sins to the Nicaean soldiers, repeating the basic principle of the imperial ideology that the empire has been granted by God as an image of his rule.[15] This act contradicts canon law and is reminiscent

[12] *Menander Rhetor*, D. Rusell – N. Wilson (eds.), (Oxford 1981), 84–86; Joseph the Philosopher, *Summation of Rhetoric*, in Ch.Waltz (ed.), *Rhetores Graeci*, 9 vols. (Stutgard/Tubingen, 1832–1836), III, 524; Angelov, *Imperial Ideology*, 82.

[13] For a detailed examination of the ideological reactions of the Nicaeans to the capture of Constantinople by the Crusaders see D. Angelov, "Byzantine Ideological Reactions to the Latin Conquest of Constantinople," in A. Laiou (ed.), *Urbs Capta. The Fourth Crusade and its Consequences* (Paris, 2005), 293–310.

[14] See Angelov, "Reactions," 293–298.

[15] N. Oikonomides, "Cinq actes inédites du patriarche Michel Autoreianos," *REB* 25 (1967), 117–118.

of the well-known suggestion of the emperor Nikephoros II Phokas that all soldiers who fall in battle should be proclaimed martyrs.[16]

While Nikephoros II's proposal had been rejected by the patriarch of the time, remission in the thirteen-century was granted by the patriarch. Both acts reflect the militaristic climate of the two periods, as well as the composition of the army in each. Nikephoros Phokas' proposal refers to a largely native Byzantine army fighting against Muslim Arabs at a time when crusading ideas were still unknown. Autoreianos' grant of remission of sin is related to the fact that a significant part of the armies of Theodore I were Latin mercenaries, who had abandoned the crusading armies and joined the forces of Nicaea and Epiros, hoping to receive better pay. It is not surprising that they had been excommunicated by the pope, although in 1211 they were hired and used by the Nicaeans against the Seljuks, with whom the Latin empire of Constantinople had formed an alliance against the Nicaeans.[17] Furthermore, in his panegyrics Niketas Choniates presents the battle of Antioch-on-the Meander, in which Theodore I defeated the Seljuks, as a clash between Christianity and Islam, and praises the emperor for having ordered his troops to carry the sign of the cross. This can be explained by the large numbers of Latins serving in his army.[18] Nonetheless, the reliance of Theodore I's army on Western European soldiers did not prevent him from promoting himself as the liberator of Byzantine cities which had been captured by the Latins.[19] That Theodore I employed Western European soldiers to fight wars of re-conquest against the Crusaders and that the Pope excommunicated the Latin soldiers who joined the armies of Nicaea and Epiros, while he did not oppose the alliance between Henry of Flanders (1206–1216)

---

[16] For Nikephoros II's proposal and for a discussion of the relative sources see A. Kolia-Dermitzaki, *Ο Βυζαντινός Ιερός Πόλεμος* [*The Byzantine Holy War*] (Athens, 1991), 137.

[17] Oikonomides, "Cinq actes," 118. In 1210 Pope Innocent III advised the Latin emperor of Constantinople, Henry of Flanders to increase the salaries of his troops in order to prevent them from joining Laskaris who was able to pay them higher salaries than the Latin emperor: *PL* 216, cols. 353–354.

[18] Choniates, *Orationes*, 175; G. Prinzig, "Der Brief Kaiser Heinrichs von Konstantinopel vom 13. Januar 1212," *B* 43 (1973), 428; R. Macrides, "From the Komnenoi to the Palaiologoi. Imperial Models in Decline and Exile," in P. Magdalino (ed.), *New Constantines. The Rhythm of Imperial Renewal in Byzantium. 4th–13th Centuries* (Aldershot, 1994), 280.

[19] In 1212, in a letter addressed to the pope the Latin emperor of Constantinople Henry of Flanders (1206–1216) remarks that Laskaris declared that he would liberate the whole of 'Greece' from the Latins: Prinzig, "Brief," 415.

and the Seljuk sultan Kaikhusraw I (1205–1211), indicates the complexity of the political and military situation of the period. Moreover, it shows that the motivation of individual soldiers was often irrelevant to the official justifications of war promoted by political and religious authorities. Many common soldiers were either unable or unwilling to understand the justifications of war and the ideals promoted by higher political and ecclesiastical authorities and ignored them. For these soldiers, material profit seems to have been a far more important incentive than any form of remission of sins granted either by the pope or the Nicaean patriarch. It should be added that in the 1220s the Seljuks hired Western European mercenaries who had previously been employed by the Nicaeans.[20]

The cultivation of an ideology of relentless militarism is reflected in the imperial encomia of the period, which emphasise the bravery and valour of the Laskarid rulers on the battlefield. The imperial orations that Niketas Choniates compiled for Theodore I underline the role of the emperor as an inspirational military leader and present him as a daring warrior. Theodore I is praised for taking risks in battle and attacking the enemy first, before his soldiers, thus setting an courageous example to his troops. He is also praised for transforming the Byzantines from cowards to fighters, from unarmed to heavily armed soldiers, from people who used to live under a roof to people who live in tents, and from men ignorant about mounting a horse to soldiers who know how to handle the armour and equipment of their horses.[21] Moreover, imperial propaganda portrays Theodore I as the killer of Kaikhusraw I in the battle of Antioch in 1211. The death of the sultan reversed the outcome of the battle when the Seljuk army fled and the Nicaeans won an unexpected victory, important for the survival of their state.[22]

Theodore I Laskaris' successors, John III Vatatzes and Theodore II Laskaris followed a more modest style of rule, in response to the limited resources of Nicaea.[23] In addition, the glorification of war

[20] A. Bombaci, "The Army of the Seljuqs of Rum," *Istituto Orientale di Napoli, Annali*, 38 (1978), 358.

[21] Choniates, *Orationes*, 123, 133.

[22] Akropolites, I, 16–17; Gregoras, I, 20–21; Macrides (*The History*, 36–38) shows that Akropolites' language in describing this event differs from the language used in the rest of his work. This reflects his reliance on another source.

[23] See Macrides, "From the Komnenoi to the Palaiologoi," 280–282.

as a holy activity disappeared from Nicaean military propaganda. Nonetheless, the main military aim of the Nicaean rulers, the recovery of Constantinople remained unaltered. After 1235, John III was able to repeatedly besiege Constantinople as a way of fulfilling Nicaea's ideological raison d'être.[24] The campaigns that he led against the Latins are promoted in the panegyrics compiled by Akropolites and Theodore II as wars for the liberation of Byzantine cities and as just punishment for the injustices of the Latins against the Byzantines.[25] As Akropolites writes in his funeral oration for John III, the emperor accepted the war against the Latins because there was nothing to gain from peace. John III is portrayed as a great general who commanded military campaigns tirelessly by night and fought without any defensive armor, holding only a sword. Jacob of Bulgaria calls him 'the best at handling the spear' and remarks that he was 'riding always the best horse.'[26] Theodore II is not attested to have led military operations against the Latin empire of Constantinople. Nonetheless, his writings emphasise the idea that the emperor should be a military leader. Theodore II considered that as emperor he should spend a large part of his day dealing with the military affairs of the empire, inspecting the army and making plans for future campaigns. He significantly increased the size of his army, which he called a 'mobile city which guards all the Roman cities.'[27] Similarly, in his epitaph on the German emperor Frederick II Hohenstaufen (1194–1250), which is mainly a treatise on kingship, he emphasises the image of the emperor as a military commander. In this work Laskaris speaks of military command as an integral part of kingship and remarks that valour, trophies, military victories, as well as suffering the difficulties of campaigning, are imperial tasks.[28]

---

[24] See J. Langdon, *Byzantium's last Imperial Offensive in Asia Minor* (New Rochelle, 1992), 82; Macrides, *The History*, 100.

[25] Akropolites, II, 18; Theodore II Laskaris, *Encomio dell' imperatore Giovanni Duca*, L. Tartaglia (ed.), (Naples, 1990), 49.

[26] G. Mercati, *Collectanea Byzantina* (Bari, 1970), I, 88; Akropolites, II, 28.

[27] Theodore II, *Epistulae*, 58; *Synopsis Chronike*, K. Sathas, (ed.), *Μεσαιωνική Βιβλιοθήκη* [*Mesaionike Vivliotheke*], 7 vols. (Venice, 1872–1894), VII, 525.

[28] Theodore II Laskaris, *Opuscula rhetorica*, L. Tartaglia (ed.), (Munich, 2000), 86–87; Angelov, *Imperial Ideology*, 252.

Changes in the military profile of the emperor after the recovery of Constantinople

With the recovery of Constantinople in 1261 and the accession of Michael VIII Palaiologos the militaristic ideology of Nicaea faded. Instead, imperial propaganda revived Komnenian models and traditional ideological claims of ecumenism.[29] This is reflected in the panegyrics compiled by the rhetorician Manuel Holobolos, whom in 1265 the patriarch Germanos appointed 'Master of the Rhetors.'[30] Unlike his Laskarid predecessors, Michael VIII is not portrayed as a heroic general who led large armies on the field. Holobolos writes that Michael VIII won victories through his piety and love of honour and not through the power of arms and the multitude of soldiers.[31] This statement vividly contrasts with Choniates' panegyrics for Theodore I and with Theodore II's attempts to increase the size of the army and his views about the role of the emperor as general. In another panegyric which Holobolos compiled for the Byzantine victory at Pelagonia against the armies of the Latin principality of Achaia, the Epirots and their German allies (1259) and for the capture of Constantinople, he emphasises the bloodless recovery of the Byzantine capital much more that the bravery on the battlefield.[32] In another imperial panegyric the patriarch Gregory of Cyprus, praises Michael VIII for his skills in handling the shield, brandishing the spear, and wielding the bow and the sword. However, he points out that the achievements and prudent policies of Michael VIII rendered warfare unnecessary.[33]

Despite the abandonment of Nicaea's militaristic ideology, Michael VIII followed an expansionist military policy which aimed at the incorporation of Epiros and Thessaly in the restored Byzantine empire and at the recovery of the Peloponnese from the Latin principality of Achaia. Michael Palaiologos' aggressive military strategy is illustrated clearly by the almost simultaneous involvement of the

[29] For the revival of Komnenian models of power under Michael VIII see Macrides "From the Komnenoi to the Palaiologoi," 273–282; Angelov, *Imperial Ideology*, 105–115.

[30] Pachymeres, II, 369. For the context of Holobolos' orations, their allusions to actual historical facts and on the title "Master of the Rhetors," see R. Macrides, "The New Constantine and the new Constantinople-1261?," *BMGS* 6 (1980), 15–41.

[31] Holobolos, *Orationes*, 38.

[32] Holobolos, *Orationes*, 70; Angelov, *Imperial Ideology*, 101.

[33] *AG*, I, 326–327, 350.

Byzantine armies in aggressive operations on four different fronts. For instance, in 1263 the forces of John Palaiologos (Michael VIII's brother) were dispatched temporarily from Macedonia and Epiros to Asia, while another army was fighting in the Peloponnese and two other armies were invading Bulgaria.[34] Probably, the policy of expansion in Europe served the interests of the imperial clan, as well as of the aristocratic groups thanks to which Michael VIII ascended the throne. The expansion in Europe could provide the group of high aristocrats who supported Michael VIII and the imperial family with the chance to accumulate wealth by exploiting the resources of the newly captured territories.[35]

As regards the ideological justification of Michael VIII's wars, the revival of ideas of imperial ecumenism enabled him to justify offensive military operations as wars aiming at the restoration of justice and imperial rule over territories which the Byzantines claimed as legitimately parts of their empire. Their political ideology supported the claim of world supremacy and of the preservation of imperial rule over the Christian *oikoumene*, which, as Metochites comments, was much larger than the territories controlled by the Byzantine state.[36] From the perspective of a territorially reduced empire, wars were justified as necessary means to restore imperial rule on lands which had been taken away from Byzantium.[37] In addition, it is logical to conclude that the emperor, who after the recovery of Constantinople in 1261 styled himself the New Constantine, would promote the restoration of the territorial integrity of the Byzantine empire.[38]

That Michael VIII advertised his wars as an attempt to restore imperial authority and portrayed himself as liberator of imperial lands is clear from his imperial propaganda. The patriarch Gregory of Cyprus sees Michael VIII's wars in the west as an attempt to get back what belonged to him. As he writes,

[34] Pachymeres, I, 285. For John Palaiologos see *PLP*, 21487.

[35] See D. Kyritses, "Κράτος και αριστοκρατία στην εποχή του Ανδρονίκου Β΄. Το αδιέξοδο της στασιμότητας," [State and Aristocracy in the Time of Andronikos II. The Stalemate of Stagnation], in L. Maurommatis (ed.), *Ο Μανουήλ Πανσέληνος και η εποχή του* [*Manuel Panselenos and his Time*] (Athens, 1999), 181.

[36] Metochites, *Miscellanea*, 717.

[37] See I. Stouraitis, *Krieg und Frieden in der politischen und ideologischen Wahrnehmug in Byzanz (7.–11. Jahrhundert)* (Vienna, 2009), 201–208.

[38] For the adoption of the epithet New Constantine by Michael VIII see Macrides, "New Constantine," 13–41.

> You see the Peloponnese, Euboia, and Attica, the ancient Greece from there you turn your eye to the Bulgarians. You know that part of our lands is occupied by them. Then you turn again your eyes to Asia. You see Ismael suppressing Israel... You could not bear your inheritance to be occupied by others.[39]

That Gregory's statements reflect the official military ideology of the period is confirmed by the Palaiologoi themselves. In his autobiography, Michael VIII states that in 1259 in the battle of Pelagonia, he defeated the 'Romans who had defected long ago.'[40] In the *typikon* for the monastery of the archangel Michael, Michael VIII refers to 'the terrible raging against us of the renegades who are of the same Roman race as we.'[41] In 1262, John Palaiologos campaigned against Michael II Angelos (1231–1271), the ruler of Epiros. He justified this campaign by stating that the despot, Michael II Angelos, could no longer claim that he had the right to occupy imperial lands, using the justification that the emperor was outside of the *patris* (Constantinople), since the emperor was restored to Constantinople. Pachymeres also relates Michael II Angelos' counter-arguments. The latter wonders how he could accept the surrender of the land his parents had gained with so much pain. He also argues that the Italians (meaning the armies of the Fourth Crusade) and not the Byzantines should claim his land, since his ancestors had seized it from them and not from the emperor in Constantinople.[42]

The *Chronicle of the Morea* is another source which reflects that the campaigns of Michael VIII on Europe were promoted as wars of recovery. According to the anonymous author, after the Byzantine victory in Pelagonia, Michael Palaiologos told his noble captives, including the prince of Achaia, William II of Villehardouin (1246–1278), 'You could not stop me from taking back the Morea which is my ancestral inheritance.' He told the Frankish knights to return to France and buy lands to support themselves and their descendants. The prince replied that the Morea was captured by nobles from *Frankia* following his father as friends and comrades. They had seized Morea by the sword and they

[39] *AG*, I, 342–343.

[40] *Imperatoris Michaelis Palaeologi de vita sua*, H. Grégoire (ed.), *B* 29–30 (1959–1960), 455.

[41] J. Thomas – A.C. Hero (eds.), *Byzantine Monastic Foundation Documents*, 5 vols. (Washington, 2000), III, 1233.

[42] Pachymeres, I, 275.

shared it among themselves and decided to make the prince's ancestors their leaders. William II concludes that the customs of his people prevent him from ceding Morea to the emperor. When a few years later the Franks defeated the Byzantines in the Peloponnese (1263), the Byzantine commanders who had been taken as prisoners told their captors that Morea did not belong to the Latins by lawful inheritance. Instead, the Franks had captured and taken possession arbitrarily of the parental inheritance of the emperor.[43]

The principle of fighting wars for the recovery of imperial territory was compatible with the Byzantine ideas of the Just War as expressed in Byzantine legal texts of earlier periods of Byzantine history and have been analysed by Angeliki Laiou. For instance, according to the *Eisagoge*, which was promulgated by the emperors Basil I, (867–886), Leo VI (886–912) and Alexander (912–913),

> The purpose of the emperor is to safeguard and maintain through his virtue the things which exist; to acquire through vigilance the things lost; and to recover through his wisdom and through just victories the things [which are] absent.[44]

Although imperial propaganda intensively promoted the idea that the empire was fighting Just Wars on its western frontiers, the shift of the empire's centre of gravity which followed the capture of Constantinople was not enthusiastically accepted by everyone. According to Pachymeres, upon hearing of the capture of Constantinople in 1261 by the Nicaeans, the *protasekrites* Michael Senachereim exclaimed 'Let nobody hope for anything good, since Romans are again walking in the city.'[45] If this statement is true and not simply a fabrication of the author in order to amplify his criticism of Michael VIII's policies with regard to Asia Minor, it indicates that even before 1261 there were people who felt that they had nothing to gain from the expansion of the empire in the west. For Pachymeres the recovery of Constantinople in 1261 forced the empire to turn westwards and the absence of the imperial presence in Asia Minor created ripe conditions for corruption among local officials and governors. Pachymeres blames the emperor

---

[43] *Chronicle of the Morea*, vv. 4244–4250, 5529–5558.

[44] *JGR*, II, Eisagoge II, 1.2; A. Laiou, "On Just war in Byzantium," in J. Langdon (ed.), *TO ΕΛΛΗΝΙΚΟΝ. Studies in Honor of Speros Vryonis, Jr.*, 2 vols. (New York, 1993), I, 163, whose translation this is.

[45] Pachymeres, I, 205.

for overtaxing the eastern provinces of the empire and for assigning unworthy and corrupt officials with the task of assessing and collecting taxes.[46] He saw the despatch of military forces from Asia Minor to Europe as a detrimental policy which denuded Asia Minor in defending itself against the Turcoman chiefdoms. Moreover, Pachymeres blames Michael VIII for not adhering to the policy of fighting on one front, which was followed by the Laskarids, and for overextending the resources of the empire in wars in Europe while Asia Minor was under intense pressure by the raids of the Turks. As he remarks, the reason why the Byzantine areas around Neokastra, the Meander, the Cayster, and Caria were deserted was that at the same time (early 1260s) John Palaiologos, was leading in the west a large and 'admirable' force made up mainly of troops from Asia Minor.[47]

The impression given by Pachymeres' narrative is that Michael VIII used Asia Minor to exact taxes and manpower in order to implement his military policies in Europe. Nonetheless, that neither Pachymeres nor any other source makes any reference to any group of leading aristocrats opposing Michael VIII's policies indicates that their interests lay in the expansion in the west and not in Asia. After the re-imposition of imperial control in the Balkans, magnates and collaborators of Michael VIII received significant grants of land in the newly annexed territories and settled there.[48] The ones who seem to have been disaffected were the lower-ranking officials such as Senachereim (see above) and the local soldiers who were anxious about their grants, privileges and properties, which were threatened by the raids and advance of the armies of the Turcoman principalities. They became prone to revolt and the throne started to treat them and their leaders with suspicion. This is indicated by the episode of the general John Doukas Angelos, who in the 1270s won significant victories against the Turks in Bithynia. In 1280, he was accused of talking with contempt about Michael VIII's second son, Constantine, and was blinded. At the same time the commander of Herakleia on the Black Sea, Michael Strategopoulos was recalled and arrested under the accusation of plotting against the emperor.[49]

---

[46] Pachymeres, I, 291–293, II, 405–407, 633–635.
[47] Pachymeres, II, 403–405.
[48] D. Kyritses, "The 'Common Chrysobulls' of Cities," *VyzSym* 13 (1999), 241.
[49] Pachymeres, II, 613–617, 621–623. John Doukas Angelos was a son of the ruler

Andronikos II and the ideological importance of defensive wars against the Turks

In the early years of his reign Andronikos II made the defence against the Turcoman chiefdoms in Asia Minor the main priority of the military policies of the empire. Andronikos II together with his father campaigned in 1280–1282 in Asia Minor and from 1290 until 1293 he settled together with his court in the area. His activities during his stay in Asia Minor were limited to repairing fortifications and building new ones. Probably, there was no large-scale fighting between the Byzantines and the Turks and the emperor's aim was through his presence to secure the loyalty of the local population and reinforce the defence of the area. In his imperial orations the future *megas logothetes* and close associate of the emperor, Theodore Metochites, comments that Andronikos II secured the safety of the population of Bithynia by reinforcing its defence through the repair of old fortifications and the construction of new ones. He makes no reference to military engagements.[50]

Despite their limited scope, the defensive operations against the Turks became ideologically important under Andronikos II. In his second imperial oration, delivered in Nicaea, where Andronikos II had taken accommodation after inspecting the frontier along the Sangarios river, Metochites describes at length the military operations of Andronikos II in Asia Minor, both those he led during the reign of his father in the early 1280s and those he conducted in the 1290s. Metochites comments that while Michael VIII was still alive the eastern provinces were in bad shape and that although almost the entire army had been transferred to the west, Andronikos II inflicted significant defeats on the Turks and restored order in Asia Minor, in particularly the areas around the Meander river and Philadelphia (modern Alaşehir). Metochites adds that Andronikos II did not change his priorities and remained focused on his aims in Asia Minor. As he writes,

---

of Epiros, Michael II Angelos. He had been brought up in the imperial court in Constantinople: *PLP*, 205; D. Polemis, *The Doukai: A Contribution to Byzantine Prosopography* (London, 1968), 95. For Michael Strategopoulos see *PLP*, 26898.

[50] Metochites, *Orations*, 364–370.

> It did not happen in the way that happens to some when the enemy attacks them from the rear during a battle, so they get distracted and immediately abandon their frontline without a defence, allowing the enemy to take advantage of the situation.

Consequently, Andronikos II's consistent and systematic efforts in Asia Minor saved it from the Turkish advance and prevented the loss of what Metochites calls, 'the best part of the empire.' Moreover, Metochites argues that before Andronikos II's intervention in Asia Minor, no one doubted that everything was lost for the Byzantines. Everyone maintained that the affairs of the Byzantines lay in shambles and it was impossible to resist on all sides. But either they could lose both parts of the empire if the resources of the state were divided between them—for it was impossible for them to address the situation on an adequate basis—or it was necessary at all costs for them to leave one part to follow its fate in a tolerable manner. The activities of the emperor on both the western and eastern fronts did not confirm these expectations.[51] Therefore, like Pachymeres, Metochites is critical of Michael VIII's insistence on expanding in Europe without having secured the eastern frontier. Andronikos II's commitment in Asia Minor is praised by Metochites' rival and predecessor in the office of *megas logothetes*, Nikephoros Choumnos. Choumnos implies that before Andronikos II's accession the defence of the east had been neglected since Michael VIII had had to face the threat posed by the king of Sicily Charles I of Anjou. Nonetheless, despite the fact that Andronikos II had to combat 'numerous enemies' in the west, he did not prove himself inferior to his father and inflicted significant defeats on the Turks.[52]

Moreover, in the imperial orations to Andronikos II the reconstruction of Tralleis is seen as a great achievement and its destruction by the forces of the Turcoman principality of Menteşe in 1283/1284 is omitted. Gregory of Cyprus points out that the rebuilding of Tralleis and the reinforcement of the fortifications of Asia Minor carried out by Andronikos II were motivated not by vanity but by material considerations and the need to protect Byzantine territories.[53] The cam-

---

[51] Metochites, *Orations*, 320–328, 338–346.

[52] *AG*, II, 25–28, 31.For the dating of Choumnos imperial oration and his career see J. Verpeaux, *Nicéphore Choumnos. Homme d' état et humaniste Byzantin (ca. 1250/1255–1327)* (Paris, 1959), 27–62.

[53] *AG*, I, 374–384. For the dating of the restoration and destruction of Tralleis see

paigns that the emperor undertook in 1280/82 and his activities during the period 1290–1293 are described as military triumphs. Despite the purely defensive nature and limited scope of the Byzantines' military operations in Asia Minor, Andronikos II is repeatedly praised for massacring the 'barbarians,' for shedding endless streams of Turkish blood and for inventing various ways to thwart, fight, and massacre the Turks. Gregory of Cuprus remarks that as a result of Andronikos II's victories the river Cayster turned red with Turkish blood. The orator Nikolaos Lampenos, who compiled his encomium between 1296–1303, writes that Andronikos II inflicted a just punishment on the Turks who no longer dare to cross the Byzantine frontier. In Lydia, as Lampenos writes, the wise leadership of the emperor led to a great victory against the Turks, who being struck by fear ran for their lives. Many of them were killed by the sword and their blood flows like a river.[54] Lampenos exclaims, 'now the Turks have recognised their master' and they are bound by oaths to serve the emperor forever.[55]

The gap between rhetoric and reality with regard to the military activities of Andronikos II is obvious. While the emperor is praised for crushing the Turks and reinforcing the defences of Asia Minor, the expansion of the Turcoman principalities and the defeats suffered by the Byzantine armies in the west at the hands of the Epirots and the Serbians caused conflicts between the throne and the aristocracy, as well as the dissatisfaction of soldiers and their leaders, whose estates and sources of income were threatened by the expansion of Byzantium's enemies.[56] Territorial reductions and the loss of state income meant that the throne lacked the resources to distribute offices and privileges to secure the loyalty of every member of the aristocracy. As a result, Andronikos II faced a considerable number of disaffected generals and

A. Failler, "La restauration et la chute definitive de Tralleis au 13e siècle," *REB* 42 (1984), 249–263.

[54] *AG*, I, 378, II, 21; Lampenos, *Encomium*, 46–48.

[55] Lampenos, *Encomium*, 51.

[56] For the Serbian advance at the expense of Byzantium in the 1280s and 1290s see L. Mavromatis, *La fondation de l' Empire Serbe. Le Kralj Milutin*, (Thessalonica, 1978), 15–53. In 1292 a Byzantine army was defeated in Epiros and the Serbians captured Skopje: *Chronicle of the Morea*, vv. 8782–9235; *Libro de los Fechos*, A.M. Fatio (ed.), (Geneva, 1885), ch. 456–468; *Livre de la conqueste de la Princée de l Amorée*, J. Longnon (ed.), (Paris, 1911), ch. 607–648; A. Laiou, *Constantinople and the Latins. The Foreign Policy of Andronikos II* (Cambridge, MA, 1972), 40; D. Nicol, *The Despotate of Epiros. 1267–1479. A Contribution to the History of Greece* (Cambridge, 1984), 38–43.

had to deal with revolts. During Andronikos II' stay in Asia Minor, his younger brother Constantine and Michael Strategopoulos were convicted of plotting to overthrow the emperor. In early 1293, they were incarcerated and were never again released.[57] Both were leading Byzantine commanders and although Pachymeres does not comment on their role in the military operations in the east, Gregoras remarks that their removal had a detrimental effect on the defence of Asia Minor.[58] In 1295, the initially successful campaign of Alexios Philanthropenos against the Turcomans in Asia Minor ended with his revolt and arrest. Philanthropenos attracted the support of the local soldiers who, according to Pachymeres, believed that while they were fighting the Constantinopolitan elite were enjoying the fruits of their struggle and leading a luxurious life, without satisfying even the basic needs of the soldiers.[59] In 1320–1321, the Byzantine army under the command of Andronikos Asan, inflicted heavy defeats on the armies of the Latin principality of Achaia and captured Akova, Karytaina and other lesser Frankish strongholds.[60] Andronikos Asan did not revolt. However, he did not conceal his dissatisfaction for not receiving an appointment that he considered appropriate for his deeds.[61]

Furthermore, local governors of Byzantine cities and provinces in Asia Minor had started co-operating with the leaders of Turcoman principalities, with whom they formed marriage alliances. In 1299, Orhan (1326–1362), the son and successor of Osman (1288–1326), was married to the daughter of the governor of a Byzantine city.[62] In

---

[57] Pachymeres, III, 171–183; For the dating of the incarceration of Constantine Palaiologos see A. Failler, "Chronologie et composition dans l'Histoire de George Pachymérès (livres VII–XII)," *REB* 48 (1990), 17.

[58] Gregoras, I, 195.

[59] Pachymeres, III, 241; See A. Laiou, "Some Observations on Alexios Philanthropenos and Maximos Planoudes," *BMGS* 4 (1978), 89–99.

[60] Gregoras, I, 297, 362–363; *Die Byzantinischen Kleinchroniken*, 3 vols. P. Schreiner (ed.), (Vienna, 1975–1979), I, 242; D. Zakythinos, *Le despotat grec de Morée*, 2 vols. (Paris, 1932–1953 repr. London, 1975), II, 71–73; A. Bon, *La Morée franque (1205–1430)* (Paris, 1969), 202; T. Shawcross, *The Chronicle of Morea. Historiography in Crusader Greece* (Oxford, 2009), 244–245. Andronikos Asan was the son of the Bulgarian king John III Asan and of Eirene Palaiologina who was Andronikos II's sister: See A. Papadopoulos, *Versuch einer Genealogie der Palaiologen. 1259–1453* (Munich, 1938), 28; E. Trapp, "Beiträge zur Genealogie der Asanen in Byzanz," *JÖB* 25 (1976), 167; *PLP*, 1489.

[61] Gregoras, I, 363.

[62] A. Bryer, "Greek historians on the Turks: the Case of the First Byzantine-Ottoman Marriage," in R.H.C. Davis – J.M.H. Wallace-Hadrill (eds.), *The Writing of History in the Middle Ages. Essays Presented to Richard William Southern* (Oxford, 1981), 471–493.

1306, the general Kassianos, an in-law of Andronikos II who was governor of the Byzantine fortresses in Mesothynia revolted. He was also accused of planning a marriage alliance with Osman.[63] Moreover, bishop Matthew of Ephesos accuses the governor of Philadelphia Manuel Tagaris of co-operating with the Turks and of marrying his daughter to a Turcoman leader.[64] Therefore, since the last years of the thirteenth century, local governors who felt that the Byzantine government could not serve their interests had been ready to join the enemy. They expected to receive a prominent position and rewards under the new regime. Kassianos' and Tagaris' intention to form an alliance with the Turks shows that joining an expanding and victorious enemy was an attractive option for dissatisfied soldiers and generals. It also indicates that knowing the weakness and inability of the centre to impose its authority over remote frontier regions, local generals were able to blackmail the throne by threatening to co-operate with the enemy.

It is reasonable to conclude that the portrayal of Andronikos II as a dedicated warrior who achieved great victories against the Turks is a response to the criticism he must have received for failing to repel the Turkish attacks and avert the Serbian advance in Macedonia and Epiros. The most severe criticism directed against Andronikos II was related to the decision to reduce the imperial fleet in 1285. Gregoras, as a supporter of Andronikos II, partly justifies this decision by claiming that there was much need of money for the defence of the land frontiers.[65] However, other contemporary authors are very critical of the decision of Andronikos II and his council to have a smaller fleet. While Michael VIII is praised for maintaining a strong fleet which achieved significant victories against the western enemies of Byzantium, recovered many islands in the Aegean Sea, supported military operations in southern Greece and reduced piracy, the decision to reduce the fleet has been seen as contributing to the loss of Byzantine control over the Aegean and of the increasing economic dependence on foreign maritime powers.[66] Consequently, it is likely that by promoting himself

[63] Pachymeres, IV, 681. For Kassianos see *PLP*, 11346.

[64] Matthew of Ephesos, *Briefe*, 106–107. For Manuel Tagaris see *PLP*, 27400.

[65] Pachymeres, III, 121; Gregoras, I, 174–175.

[66] Pachymeres, III, 81–83, IV 627; Gregoras, I, 174, 207; G. Westernik, "Le basilikos de Maxime Planoude," *BSl* 28 (1967), 65–66; H. Ahrweiler, *Byzance et la mer* (Paris, 1966), 369; Laiou, *Constantinople and the Latins*, 74–75; S. Lampakes, *Γεώργιος Παχυμέρης. Πρωτέκδικος και δικαιοφύλαξ* [*George Pachymeres. Protekdikos and dikaiophylax*] (Athens, 2004), 103.

as a victorious general Andronikos II wished to stave off criticism, to assert his authority over the populations of Asia Minor, to display his determination to defend the empire's eastern frontier and to deter separatist movements initiated by local governors who were tempted to join the enemy. However, the local soldiers in Asia Minor were already alienated by the policies of Michael VIII and suffered from the consequences of the raids and advance of the Turcoman emirates in the east and of the Serbians in the west. Having lost their incomes and sources of wealth they became prone to rebellion, particularly since the imperial army failed repeatedly to gain any decisive victory on the battlefield. The leadership for the rebellions was provided not by local magnates but by successful generals, such as John Doukas Angelos, Alexios Philanthropenos and Kassianos who were members of the aristocratic elite and the imperial family. They had little or no connection with the regions where they commanded, but their military reputation and prowess were sufficient to attract the support of the local soldiery.

The military profile of the emperor under Andronikos III and John VI Kantakouzenos

Unlike his grandfather, Andronikos III did not make extensive use of imperial propaganda to promote himself as a great warrior and justify his military policies. He was the recipient of a single panegyric, which makes brief reference to his military deeds against Turcoman raiding parties in Thrace.[67] This reflects the style of government of Andronikos III. As Gregoras remarks Andronikos III introduced an informal style of government, probably to distinguish himself from his grandfather, whom he dethroned in 1328.[68] Most of our knowledge of Andronikos III's military policies derives from the *Histories* of John Kantakouzenos. Being the emperor's closest associate, Kantakouzenos depicts Andronikos III as an excellent soldier and strategist. By presenting him as a vigorous and great warrior, Kantakouzenos criticises Andronikos II, who was not a soldier emperor, and justifies his support for the younger Andronikos during the civil war of 1321–1328. For instance, describing a Bulgarian raid in 1328, Kantakouzenos

[67] Gregoras, *Logos*, 23.
[68] Gregoras, I, 565–566; Angelov, *Imperial Ideology*, 47.

claims that during Andronikos II's reign the Bulgarians were able to invade Byzantine lands without resistance, because there was no one to confront them. Instead, Andronikos II was always trying to resolve the differences with them through embassies and promises. Because of this, as Kantakouzenos writes, the Bulgarians believed that the Byzantines would not confront them. Now the Bulgarian king, Michael Šišman (1323–1330), seeing the large and efficient army assembled by Andronikos III, changed his mind and decided to retreat.[69]

For Kantakouzenos, that the emperor should be a competent soldier was an essential requirement for his elevation to the throne. Among the reasons he gives in not supporting Andronikos II's inclusion of Michael Katharos, the illegitimate son of Constantine Palaiologos (Andronikos II's second son), in the line of imperial succession after the death of the co-emperor Michael IX in 1320, was that Katharos did not have any military training and experience.[70] However, the real motive behind Kantakouzenos' statement was his opposition to the removal of Andronikos III from the line of succession. Nevertheless, this statement reflects his perceptions concerning the role of the emperor as military commander. In another instance, Kantakouzenos states that when Andronikos III fell seriously ill (1329), the empire needed to be led by someone who was not only an experienced and good general but also a great general, who was very well trained in matters of war, obviously implying himself.[71] In the same context, Kantakouzenos reports that Alexios Apokaukos had told Andronikos Asan that he was the proper person to succeed Andronikos III, 'not only because of his bravery and experience in wars, but also due to his illustrious ancestry and prudence.'[72]

Under Andronikos III and John Kantakouzenos, the wars against the Turks and the defence of Asia Minor lost their ideological importance. This reflects the changes at the time. After leading a campaign against Orhan in Bithynia in 1329, which ended with the Byzantine defeat at the hands of the Ottomans in Pelekanos-Philokrene, Andronikos III followed a policy of containment towards the Turks. In 1333, he agreed to pay an annual tribute of 12,000 *hyperpyra* to Orhan for the protection of the fortresses of Mesothynia which were under Byzantine

[69] Kantakouzenos, I, 328.
[70] Kantakouzenos, I, 15.
[71] Kantakouzenos, I, 425.
[72] Kantakouzenos, II, 115.

control.[73] Instead, the reign of Andronikos III saw the revival of the ideology of war to recover territories which had belonged to the empire before 1204. Andronikos III's campaigns in Epiros, Thessaly, Chios, Lesvos and Phokaia aimed at the imposition of direct imperial rule over territories that had already recognised nominal imperial authority during the reigns of Michael VIII and Andronikos II. This policy was implemented with the significant military assistance of Turcoman chiefdoms. In 1335, when Andronikos III had secured the alliance and military aid of the emirates of Saruhan and Aydın, he attacked the Genoese lord of Phokaia, Arrigo Tartaro, who was ruling in the name of Andreolo Catanea.[74] In 1338, the reinforcements sent by Aydın played a decisive role in the Byzantine victory against the Albanians, which contributed to the imposition of direct imperial control over the state of Epiros.[75]

The ideology of fighting for the restoration of imperial authority is articulated in Kantakouzenos' account of the campaign that Andronikos III led against the Epirots in 1339/1340. Kantakouzenos writes that, during the siege of Arta, he told the defenders of the city that it was unjust to prefer to be ruled by the Tarantians (he implies Philip of Taranto) and not by the Byzantine empire, which has been their ancestral authority since the days of Caesar.[76] He also argues that the Angeloi had not gained their authority by liberating Epiros from 'barbarians.' Instead, they were subjects of the Byzantine emperors and received from them their annual administrative authority (*epitrope*) over the land.[77] However, they took advantage of the war between the Byzantines and the Latins (he means the Fourth Crusade) and usurped imperial authority over Epiros because the Byzantine emperors were cut off from these lands. Kantakouzenos concludes that when the Byzantine emperors with God's aid expelled the Latins from the Byzantine lands, they unified the parts of Europe and Asia that they

[73] *Kleinchroniken*, I, 80; Gregoras, (I, 458) refers to heavy tribute forced upon the Byzantine possessions in Bithynia.

[74] Gregoras, I, 529–532; Kantakouzenos, I, 480–481.

[75] Kantakouzenos, I, 496–498.

[76] For the connection between Philip of Taranto and Epiros see Nicol, *The Despotate of Epiros 1267–1479*, 114–115.

[77] It is likely that Kantakouzenos alludes to the fact that when Theodore Komnenos Doukas left Asia Minor for Epiros about 1207 to take over after the death of his brother he was made to swear an oath of loyalty to the emperors of Nicaea. See Akropolites, I, 24–25; Nicol, *The Despotate of Epiros 1267–1479*, 2.

had formerly ruled. When they demanded Epiros from the Angeloi, they were not given it. Therefore, the emperors were deprived unjustly of these lands and were forced to undertake many campaigns against Epiros. As Kantakouzenos told the Epirots,

> It could be said that the emperor by campaigning against you—and with God's aid he will make you his servants—is not being unjust, since he is not seizing somebody else's authority. Instead, he is doing the right thing by claiming his paternal authority. The Angeloi cannot base their authority on the succession rights of Nikephoros Angelos, because they were under the rule of the Roman Emperor for many more generations.

The outcome of this conflict was the surrender of Arta, but not because the Epirotes were persuaded by Kantakouzenos' arguments. Instead, Arta's arguments for its surrender were based on realistic calculations. It justified its support of Nikephoros Angelos by claiming that it thought it would be ingratitude of the worst kind not to assist him with all its power to defend his paternal inheritance. This statement shows that Kantakouzenos and the people of Arta rested on different definitions of the term 'ancestral authority.' According to Kantakouzenos the people of Arta decided to surrender when they realized that they would be lost along with their city. Its people requested Kantakouzenos 'to arrange our affairs in such a way as to prove that the service to the emperor is better than the freedom of the fatherland (*patris*).'[78] This statement indicates that the Epirotes did not see the Byzantine emperor as their legitimate ruler. It is worth noting that, regarding the same campaign, a contemporary patriarchical document states that the Angeloi should rule Epiros as governors (*kephalai*) and not as independent rulers.[79]

Although the military campaigns of Andronikos III were justified as attempts to restore direct imperial control over territories which were legitimate parts of the empire, Kantakouzenos' account reveals the gradual reduction of the control that the throne exerted over the army and the increasing influence of powerful aristocrats. Under Andronikos III it was not uncommon for wealthy individuals to build fortifications, to fund imperial campaigns and to contribute to the payment of soldiers. John Kantakouzenos, Alexios Apokaukos and John Vatatzes are three

---

[78] Kantakouzenos, I, 520–524.

[79] *Das Register des Patriarchats von Konstantinopel 1315–1355*, 3 vols. O. Kresten *et al.* (eds.), (Vienna, 1981–2001), II, 96.

well-known cases of powerful individuals who constructed fortresses.[80] It is probably no coincidence that such individuals were close associates of the emperor. It seems that the severe financial difficulties of the empire and the exhaustion of its resources as a result of the civil war led the emperor to allow wealthy individuals, who had gained imperial favour, to construct fortifications in Thrace which could be used for the defence of the area against a possible large-scale Serbian invasion and against Turkish raids. Instead of the state, these individuals would take up the cost of protecting their estates and the local population.

Beyond the construction of static defences, the throne lost the prerogative of maintaining and funding the imperial army and its campaigns. Kantakouzenos twice remarks that it was he and not the treasury who paid the salaries of the mercenaries.[81] Similarly, Kantakouzenos relates that in 1329 the construction of a fleet and the preparation of the campaign of Andronikos III in Chios were expensive undertakings and were funded partly by wealthy aristocrats, who probably wanted to promote their own private interests, in finance and commerce. In this instance, private interests and imperial military strategy coincided.[82] In 1340, Alexios Apokaukos contributed financially to the building up of a military fleet with the official mission of protecting the Byzantine territories from raiding by the fleets of the Turcoman principalities. In exchange for paying for the construction of the fleet, Apokaukos requested the incomes of the Byzantine islands in the Aegean. Nowhere in his account has Kantakouzenos ideologically objected to the financing of the imperial army and military campaigns by individuals. When he refers to his personal activities, he claims that by funding the imperial army he reinforced the imperial authority. In 1322 he told both emperors that he wished to be outstripped by no one in giving money and servants, as well as his services, for the benefit of people of the same ethnicity and for the honour of the emperors.[83] By claiming that he did so in the interests of the empire and for its protection, Kantakouzenos attempts to legitimise his acts.

[80] Kantakouzenos, II, 184, 475; Gregoras, II, 585, 708, 741–742.

[81] Kantakouzenos, I, 137, 184–185.

[82] Kantakouzenos, I, 381; U. Bosch, *Kaiser Andronikos III Palaiologos. Versuch einer Darstellung der Byzantinischen Geschichte in den Jahren 1321–1341* (Amsterdam, 1965), 117; K.-P. Matschke, *Fortschritt und Reaktion in Byzanz im 14. Jahrhundert: Konstantinopel in der Bürgerkriegsperiode von 1341 bis 1354* (Berlin, 1971), 234.

[83] Kantakouzenos, I, 185.

Magnates who contributed financially to the maintenance of the army and fortifications must have played an increasing role in the formation of imperial military policy. It should also be expected that troops such as those who were paid by Kantakouzenos and not the imperial treasury were primarily loyal to their paymaster and not the throne. Gregoras remarks that long before the outbreak of the civil war of 1341–1347 the Byzantine soldiers viewed Kantakouzenos as their 'wealth provider.'[84] Undoubtedly, this statement implied more than the expected courage and admiration which a good general would inspire in his troops. Nonetheless, it should be noted that people such as Kantakouzenos and Apokaukos did not intend to challenge the emperor. They aimed at increasing their influence on the throne by increasing their influence on the army and fleet of the empire.

In 1341, the unexpected death of Andronikos III was followed by the outbreak of a catastrophic civil war between Kantakouzenos, who claimed to be the rightful regent of Andronikos III's son, John V Palaiologos and the regency in Constantinople, which was led by the Alexios Apokaukos, the patriarch John Kalekas, and Andronikos III's widow, Anna of Savoy. This civil war, which ended in 1347 with the victory of Kantakouzenos' party, was a conflict among aristocratic cliques, which fought over access to the empire's limited resources through the exertion of influence on the throne. The rival parties relied on networks based on kinship, common interest and patronage. The empire came out of the civil war very much weakened. It suffered dramatic territorial loses, its manpower was reduced and the strength of its army was diminished.[85]

One of the consequences of the civil war was the weakening of central control over the military resources of the empire. The *chrysobull* which John VI Kantakouzenos issued in 1341/1342 to appoint his nephew John Angelos ruler of Thessaly provides a characteristic example.[86] John Angelos, it said, would rule Thessaly independently. His only obligation was to provide full military assistance to the emperor when he was campaigning west of Christoupolis (modern Kavala).

---

[84] Gregoras, I, 586, 614.

[85] For a detailed analysis of the civil war of 1341–1347 see Matschke, *Fortschritt und Reaktion*, passim; D. Kyritses, *The Byzantine Aristocracy in the Thirteenth and Fourteenth Centuries*, unpublished PhD diss. (Harvard, 1997), 358–377.

[86] Kantakouzenos, II, 312–322. See H. Hunger, "Urkunden–und Memoirtext: Der Chrysobullos Logos des Johannes Kantakuzenos für Johannes Angelos" *JÖB* 27 (1978), 107–125. For John Angelos see *PLP*, 204.

If the emperor would campaign east of Christoupolis, he could help the emperor only if, and to the extent, he could. Therefore, the soldiers of Thessaly would reinforce the imperial army only when it was to campaign in territories close to their home and the basis of income. Moreover, the need to accommodate the ambitions of his sons and of John V Palaiologos led Kantakouzenos to carry out the partition of the empire. Each prince was assigned a territory which he ruled independently. Matthew Kantakouzenos became co-emperor and ruled Thrace, John V was sent to Thessalonica and Manuel Kantakouzenos ruled the Morea from 1349 until his death in 1380. That from 1352 until 1357 Matthew Kantakouzenos and John V Palaiologos fought against each other proves that John VI, the senior emperor in Constantinople, had no control over their military forces.

Fighting the impious barbarians and natural enemies

In 1354, John V Palaiologos took over as sole emperor and John VI Kantakouzenos was forced to abdicate. Kantakouzenos portrays his successor as a good soldier and concludes his *History* by praising him for capturing Mesemvria from the Bulgarians. However, this is not to deny the military weakness and impoverishment of the empire. A *chrysobull* issued by John V shortly after his accession states that there was no money in the imperial treasury. Moreover, John V had to repay the loans he had received for his military effort against Kantakouzenos.[87]

The Byzantine army was unable to resist effectively the armies of the Ottoman frontier lords, who in the 1360s established themselves in Thrace and captured such important cities as Didymoteichon and Adrianople and by the 1380s, had expanded in Macedonia.[88] In the

[87] L. Perria, "Due documenti greci del XIV secolo in un codice della biblioteca vaticana (vat.gr. 1335 con un tavole)," *JÖB* 30 (1981), 263. In1343, John V's mother, Anna of Savoy had borrowed 30,000 ducats from Venice for the war effort against Kantakouzenos and in 1352 John V, who at this time was in Ainos, borrowed 5,000 ducats to fund his operations against Kantakouzenos who was then emperor in Constantinople. See P. Katsoni, *Μία επταετία κρίσιμων γεγονότων. Το Βυζάντιο στα έτη 1366–1373* [*Seven Critical Years. Byzantium 1366–1373*] (Thessalonica, 2002), 73.

[88] For the nature of the Ottoman expansion in Thrace and Macedonia in the second half of the fourteenth century see H. Lowry, *The Shaping of the Ottoman Balkans 1350–1550. The Conquest, Settlement and Infrastructural Development of Northern Greece* (Istanbul, 2008), passim.

1370s the Byzantine empire was reduced to Constantinople and its suburbs, a few cities in the Black sea and in Thrace, such as Selymvria, Raidestos, and Anhialos, and a few islands in the northern Aegean. Thessalonica was isolated and in tenuous contact with the capital, while the rulers of the Morea followed their independent foreign and military policies. The city of Philadelphia in Asia Minor, which surrendered to the Ottomans in 1392, had long since lost any contact with Constantinople. In the 1370s, the failure of the diplomatic efforts of John V to attract Western European military aid and the defeat of the Serbians at the battle of Maritza in1371 led him to seek peace with the Ottomans and become a tribute-paying vassal of Murad I (1362–1389).[89] Throughout the last century of the existence of the Byzantine empire the aim of its rulers was either to use diplomacy to attract military aid from Western Europe against the Ottoman Turks, usually in the form of a Crusade, or to appease the Ottomans.

In this context of dramatic territorial reduction and defensive warfare against the Ottomans, imperial propaganda articulated an imperial profile which emphasised the ideological importance of such defensive wars. This is reminiscent of Andronikos II's reign when wars against the Turks were ideologically important. However, the context was quite different. While in the late thirteenth century Andronikos II is praised for shedding rivers of Turkish blood in Asia Minor, fifteenth-century emperors are praised for defending the walls of Constantinople. In the 1420s the bishop of Kiev, Isidore, praises John VIII for his command during the siege of Constantinople in 1422 which was led by Murad II (1421–1451) and the frontier lord Mihaloğlu. As Isidore writes, John VIII patrolled the walls throughout the night.[90] John VIII is also praised for repairing the walls and the harbour of the Byzantine capital.[91] The central role in the defence of Constantinople according to the imperial propaganda reflects the threat that the Ottomans posed

[89] Ostrogorsky argues that Byzantium became tributary to the Ottomans immediately after the defeat of the Serbians: G. Ostrogosky, "Byzance état tributaire de l'Empire turc," *ZRVI* 5 (1958), 49–58. P. Katsoni places the event in 1376: "O χρόνος έναρξης της Βυζαντινής υποτέλειας στους Οθωμανούς," [The Year of the Beginning of the Vassalage of the Byzantines to the Ottomans] *Byzantiaka* 14 (1994), 459–481.

[90] *PP*, III, 177; For an analysis of Isidore's oration see O.J. Schmitt, "Kaiserrede und Zeitgeschichte im späten Byzanz: Ein Panegyrikos Isidoros von Kiew aus dem Jahre 1429," *JÖB* 48 (1998), 209–242. The author argues that this imperial oration does not follow the rules of the genre.

[91] *PP*, III, 297–298.

to Constantinople, which in the course of the fifteenth century had evolved into a city-state, and the fact that control over the capital was the main subject of the conflicts within the imperial family. Manuel II had clashed with his brother Andronikos IV (1376–1379) and John VII (1390–1391) (Andronikos IV's son), while John VIII had to face the ambition his brothers.[92]

Moreover, as Byzantium became increasingly threatened by the Ottomans in the second half of the fourteenth century, so its religious and cultural identity came to the fore. The struggle for survival took the form of a struggle between good and evil. The conflict with the Ottomans is described by late fourteenth and fifteenth century authors as a fight against impious barbarians. The promotion of this idea aimed at emphasising the cultural superiority of the Byzantines and at attracting military aid from Western Europe. Thus, the Byzantines promoted the notion of religion as the dividing power between 'us' and the Turks who are portrayed as a dangerous, and often uncivilised, enemy against whom the Byzantines are involved in an unavoidable war.[93] As will be seen below, this reflects not only the threat that the Ottomans posed to the very existence of the Byzantine state but also the conflicts within Byzantine society.

Kantakouzenos remarked that his mission as an emperor was to bring peace to the Christians and to thwart the 'Ismaelites.'[94] He used the impiety of the enemy as a justification of the wars against the Turks, whom he presents as 'barbarians' and 'natural enemies.' During the civil war of 1321 and 1328 Kantakouzenos mentions that he and Andronikos III regretted that the Byzantines were fighting each other. Instead, they should contemplate how they would organise the war against the barbarians—implying the Turcoman principalities—who were Byzantium's natural enemies.[95] Furthermore, Kantakouzenos relates that he was willing to offer as much help as he could in an attack

---

[92] Kiousopoulou, *Βασιλεύς ή οικονόμος*, 173. For the conflicts within the imperial family in the late fourteenth and fifteenth centuries see G. Kolias, "Η ανταρσία του Ιωάννη Ζ′ Παλαιολόγου εναντίον του Ιωάννη Ε′ Παλαιολόγου (1390)," [The Revolt of John VII Palaiologos Against John V Palaiologos (1390)] *Hellenika* 12 (1952), 34–64; J.W. Barker, *Manuel II Palaeologus (1391–1425): A Study in Late Byzantine Statemanship* (New Brunswick, 1969), 17–34; N. Necipoğlu, *Byzantium Between the Ottomans and the Latins* (Cambridge, 2009), 119–148.

[93] For similar developments in the seventh century as a result of territorial losses see R. Browning, "Greeks and Others: From Antiquity to the Renaissance," in T. Harrison (ed.), *Greeks and Barbarians* (Edinburgh, 2002), 268–269.

[94] Kantakouzenos, III, 20.

[95] Kantakouzenos, I, 160.

against the Turks organised by the pope and the Western Europeans and he would be very happy to see the Turks being punished for the sufferings they caused to the Christians.[96] In a more explicit manner, wishing to blame Alexios Apokaukos, his main enemy in the civil war of 1341–1347, for being the first to employ Turkish allies, he states that the Turks are barbarians who enjoy killing and enslaving Byzantines mercilessly, since they are their enemy by nature due to the 'extreme differences with us in piety.'[97] In this instance, Kantakouzenos, referring to the barbarians fighting for his enemies, identifies barbarism with cruelty.

Despite his anti-Turkish rhetoric and the repetition of stereotypical views about barbarians, Kantakouzenos admits that it was impossible to face the Turks militarily and continued the policy of appeasement and co-operation that he and Andronikos III had initiated in the 1330s. Moreover, Kantakouzenos appreciated the military strength of the Turks. As he writes, in 1354 he vehemently opposed the proposal put forward by younger aristocrats to face the Ottomans militarily, since,

> In terms of military experience they are not inferior to the Byzantines, but instead are far superior to them in terms of numbers, more vigorous in their preparations for war, better prepared and readier to campaign than the Byzantines.[98]

Moreover, since the 1330s Kantakouzenos had established a close relationship with the leader of the emirate of Aydın, Umur. According to the late fifteenth century work of Enveri, Umur considered Kantakouzenos his brother.[99] In the 1330s, the emirates of Aydın and Saruhan assisted the Byzantine operations in Phokaia, Chios and Lesvos and Epiros. During the civil war of 1341–1347, and after Stefan Dušan, withdrew his support, Kantakouzenos relied substantially on the military assistance of Umur and various other groups of Turks.[100] When in 1344 the Latins captured Smyrna, the emirate of Aydın lost its main naval base and strategic importance for the rival parties in the Byzantine civil war. Consequently, Kantakouzenos turned to the

---

[96] Kantakouzenos, III, 57.
[97] Kantakouzenos, II, 396.
[98] Kantakouzenos, III, 295–296.
[99] I. Melikoff-Sayar, *Le Destan d' Umur pascha*, (Paris, 1954), v. 353.
[100] Kantakouzenos, II, 344–348, 390–407, 529–534.

Ottomans and Orhan, to whom he gave his daughter Theodora in marriage in 1346.[101] Kantakouzenos' familiarity with the Turks is made clear by an incident which he describes. In 1347, he confronted a party of Turkish raiders. He claims that he knew their leaders, Karamehmed and Martuman, who were killed in the ensuing battle. Kantakouzenos relates that he asked the remaining Turks to join him, since they had repeatedly campaigned together during the civil war of 1341–1347.[102]

In addition, for Kantakouzenos this alliance with the 'barbarians' and 'impious' against the enemies of the empire was an acceptable practice. They could be used in the Just Wars fought for the recovery of imperial territory. This is the ideological justification for Kantakouzenos' decision to request military aid from the Ottomans in order to recover Byzantine cities captured by the Serbians during the civil war of 1341–1347.[103] Kantakouzenos also calls Umur, a friend (*philos*) because he supported the imperial campaigns led by Andronikos III in Phokaia and in Epiros.[104] Therefore, Kantakouzenos' rhetoric against the Turks should not be taken literally. The numerous references to 'impious barbarians' can be explained by the fact that Kantakouzenos' account promotes aristocratic attitudes toward warfare, such as fighting for faith and honour; that he was an emperor whose official task was to defend the Christian Roman empire; and that he tried to exonerate himself from the charge that he contributed to the expansion of the Turks by using them as allies and mercenaries in the civil conflicts of the 1340s and 1350s. It should not be forgotten that Kantakouzenos compiled his account in the 1360s when the Ottomans were starting to expand in Europe at the expense of Byzantium.

Demetrios Kydones consistently describes the conflict with the Ottomans in the second half of the fourteenth century as a war between the Christians and the impious barbarians. Kydones considered that there was an urgent need for Byzantium to gain what he calls 'the sympathy of the Christians,' who would carry out an attack against the Turks in Asia.[105] In 1363, in support of this argument, he argued, that the people who live in Western Europe are very willing to fight and

[101] Kantakouzenos, II, 586–589; Bryer, "Marriage," 471–493.

[102] Kantakouzenos, III, 65.

[103] Kantakouzenos, III, 111.

[104] Kantakouzenos, II, 398.

[105] Kydones, *Correspondance*, I, 125–128; F. Tinnefeld, *Demetrios Kydones Briefe*, 4 vols. (Stuttgart, 1981–2003), I, 353–362.

die for Christ.[106] In a letter written in 1385 which was addressed to the despot of the Morea, Theodore I Palaiologos (1383–1407), the brother of Manuel II, who was involved in military conflict with the Navarrese Company, Kydones describes the conflict between the Byzantines and the Ottomans as continuous and inevitable, due to the barbarity of the enemy. As he writes, the despot of the Morea was fighting against enemies with whom he shared common customs, faith and laws. However, he comments that,

> There is nothing in common between our war and your war. Our war against the barbarians is continuous, as if it is an inheritance given by nature. We are prudent, while they behave licentiously. Our main difference is in faith.[107]

Similarly, when in 1371 Murad I sent an envoy to Constantinople to negotiate the return of Gallipoli to the Ottomans, Kydones compiled a narration which sought to dissuade the people of Gallipoli from surrendering to the Turks. In this he presents the enemy as greedy barbarians and exhorts the people not to prefer slavery to freedom.[108]

It is noteworthy that Serbia and Bulgaria are not included in the alliance against the Ottomans envisaged by Kydones. In 1366, he wrote that it would be ideal to ally with the Serbians and Bulgarians against the Ottomans; however, the Byzantines should not forget that in the past the Serbians and the Bulgarians had attacked and seized Byzantine lands without any provocation, that they had shown brutality towards the Byzantines, that they had imposed heavier taxation than the Turks, and they had not accepted the Byzantine proposal for alliance based on their shared religion.[109] Kydones, who had converted to Catholicism, made this statement at a time when the priority of the Byzantine state was to attract military aid from the Latin West. Therefore, it was important that people would be convinced that, on the basis of past experience the Byzantines could not expect any aid from their Orthodox neighbours. Furthermore, in the 1360s Bulgaria and Serbia were too fragmented and too weak to offer any military assistance.

[106] G. Mercati, *Notizie a Procoro e Demetrio Cidone, Manuele Caleca e Teodoro Meliteniota* (Vatican, 1931), 372–373.

[107] Kydones, *Correspondance*, II, 239; Tinnefeld, *Briefe*, III, 227–230.

[108] *PG* 154, cols. 1010, 1020. For the dating of this oration see Necipoğlu, *Between the Ottomans and the Latins*, 123.

[109] *PG* 154, cols. 972–973.

The Ottoman Turks are identified as barbarians in a patriotic discourse compiled by Manuel II in 1383. Trying to persuade the inhabitants of Thessalonica not to surrender unconditionally to the Ottomans, Manuel II emphasises stereotyped views about the barbarians. He comments that submission to the Ottomans would be a shameful slavery to impious barbarians and that death would be preferable. He adds that the barbarians wanted to impose on Thessalonica the yoke of slavery, which is a heavy burden, as well as a shameful thing. He exhorts the people of Thessalonica to fight for their freedom and not to try to reach an agreement with the enemy, because what the barbarians see as treaties is in fact slavery. Furthermore, Manuel II remarks that it is better to fight and die gloriously than to submit to slavery. He repeats this comment in the conclusion of his oration stating that,

> You should consider death as a friend and benefactor, since you chose not to be enslaved by the barbarians, who are living like beasts and have customs of an unusual nature.[110]

In addition, in his funeral oration for his brother the despot of Morea, Theodore I, who died in 1407, Manuel states that Theodore had tried to raise the spirit of his subjects by every possible encouragement and had urged them not in any way to betray themselves to the infidels, who are totally opposed to the Byzantines in customs, religious beliefs, laws and in all other areas.[111]

The patriotic discourses of Manuel II and Kydones should not be seen exclusively as reactions to the Ottomans' expansion. They should also be examined in the context of the conflicts within Byzantine society. When Manuel II delivered his speech to the Thessalonians he had to face the hostility of a significant segment of the population, who had to bear the brunt of the cost of the defence and were in favour of an accommodation. In 1387, this party prevailed and surrendered Thessalonica to the Ottomans. As he wrote shortly afterwards, while he was risking his life there were people in Thessalonica who were fighting on the side of the enemy.[112] Moreover, the fall of Thessalonica

---

[110] Manuel II Palaiologos, *Συμβουλευτικός πρός τούς Θεσσαλονικεῖς,* [*Discourse of Counsel to the Thessalonicaeans*] V. Laourdas (ed.), in *Makedonika* 3 (1953–1955), 295–301.

[111] Manuel II, *Funeral Oration*, 127–129, 191.

[112] Manuel II, *Letters*, 187; Barker, *Manuel II*, 79–80.

meant the loss of Manuel's realm, since he had established his own independent regime in Thessalonica.[113] The attitude of the people of Thessalonica reflects that in the period of the Ottoman expansion there were limits to the incentives and ability to oppose the new regime. It has been concluded that for a major part of Macedonia the Ottoman conquest was followed by a sharp decrease of the taxes that burdened the peasantry during the Byzantine period.[114]

Although Manuel II exhorted the people of Thessalonica to prefer death to slavery he served in the Ottoman army as a vassal of Bayezid I.[115] In 1391, he followed Bayezid I on his campaign in Sivas against Burhan ed-din Ahmed and in 1392 he was on the sultan's side when Philadelphia surrendered to the Ottomans. Manuel II justified himself by stating that he was driven to follow the Sultan on his campaigns by compelling necessity which he had to obey for the sake of saving his subjects from disaster and because it would have been madness to disobey with a drawn sword brandished at him.[116] Similarly, in the speech that he delivered to protest against the cession of Gallipoli, Kydones points out the activities and influence of the people who favoured the surrender of the city and were well-disposed towards the Ottomans. Eventually, the Ottomans entered the city of Gallipoli 1377.[117] Nor did Kydones' letter to Theodore I, prevented the despot of Morea from calling on the help of the Ottoman commander Evrenos beg against his external and internal enemies.[118] Therefore, most probably the identification of the Turks with barbarism had limited appeal to the Byzantine population and the authors who repeatedly call the military conflicts with the Turks wars against 'impious barbarians' had personal motives for doing so. Moreover, social and economic developments,

[113] G.T. Dennis, *The Reign of Manuel II Palaeologus in Thessalonica, 1382–1387* (Rome, 1960), 77; Necipoğlu, *Between the Ottomans and the Latins*, 46–52.

[114] N. Oikonomides, "Ottoman Influence on Late Byzantine Fiscal Practice," *SüdostF*, 45 (1986), 23–24; P. Frankopan, "Land and Power in the Middle and Later Period," in J. Haldon (ed.), *A Social History of Byzantium* (Oxford, 2009), 120.

[115] Manuel II, *Letters*, 38; Barker, *Manuel II*, 79–80.

[116] E. Zachariadou, "Manuel II Palaeologus on the Strife between Bayezid I and Kadi Burhan al-Din Ahmad," *Bulletin of the School of Oriental and African Studies*, 43 (1980), 471–481; Manuel II, *Funeral Oration*, 135.

[117] *PG* 154, col. 1019; Necipoğlu, *Between the Ottomans and the Latins*, 123–125. For the importance of Gallipoli in the military conflicts of the period see H. İnalcık, *Gelibolu*, *EI*², cols. 983–987.

[118] See Necipoğlu, *Between the Ottomans and the Latins*, 239.

such as the presence of Byzantine merchants in Ottoman territory, as well as the long coexistence in frontier regions such as Bithynia, must have made many Byzantines familiar with the culture and way of life of the Turks. It is reasonable to conclude that for these people the enemy was not completely evil and barbarian.[119]

---

[119] See the discussion of Necipoğlu, *Between the Ottomans and the Latins*, 208.

CHAPTER TWO

COMMANDERS

The military character of the late Byzantine aristocracy

Most late Byzantine aristocrats sought a military career and regarded a career in the fiscal and civil bureaucracy as beneath their dignity.[1] High aristocrats were attracted to military office because military power brought them closer to the emperor and provided the means to secure and improve their situation vis-à-vis the throne and the imperial household. Moreover, appointments to offices or grants of land and state resources were made by the emperor and were not hereditary. This meant that it was essential for the Byzantine aristocrats to promote their interests through imperial favour.[2] In an oration compiled for John V in the mid-fourteenth century, Kydones outlines the benefits for office holders: They could acquire land and houses, lend money to merchants and increase their influence on the government by gifts from those who wished to approach the emperor for appointments to office. They could also provide their kinsmen with access to state resources.[3] It is certain that high-ranking military officers enjoyed these benefits. The career of John Kantakouzenos is the most characteristic example of an ambitious aristocrat who used his military office and the support of the army to increase his influence, secure imperial favour and in the end usurp the throne. In 1320, Kantakouzenos appears to have been an associate of the co-emperor Michael IX and in command of military forces in Gallipoli.[4] After Michael IX's death, Kantakouzenos gave financial support for the revolt of Andronikos III

---

[1] K.-P. Matschke – F. Tinnefeld, *Die Gesellschaft im späten Byzanz* (Koln/Weimar/Vienna, 2001), 20–21; D. Kyritses, "Η Άλωση της Κωνσταντινούπολης και το τέλος του Βυζαντινού πολιτισμού," [The Fall of Constantinople and the end of Byzantine Civilization], in T. Kiousopoulou (ed.), *Η Άλωση της Κωνσταντινούπολης και η μετάβαση από τους μεσαιωνικούς στους νεότερους χρόνους* [*The Fall of Constantinople and the Transition from the Medieval to the Modern Age*] (Rethymnon, 2005), 164.

[2] Frankopan, "Land and Power," 126; J. Haldon, "Social Élites, Wealth and Power," in J. Haldon, (ed.), *A Social History of Byzantium* (Oxford, 2009), 182.

[3] Kydones, *Correspondance*, I, 21.

[4] Kantakouzenos, I, 24.

and the army received many payments from him and not from the imperial treasury. Inevitably, the soldiers of the imperial army would be more loyal to Kantakouzenos than to the emperor. In the early 1320s, as a result of the agreement between Andronikos II and Andronikos III, he was promoted to the office of *megas domestikos* which he held until the death of Andronikos III in 1341. Kantakouzenos' promotion to this office meant that he became the head of the entire Byzantine army.[5] Accompanying Andronikos III on all his important campaigns enabled Kantakouzenos to increase his influence on the emperor and on the army. Indeed, during the civil war of 1341–1347, Kantakouzenos relied substantially on the support of the army. Immediately after the death of Andronikos III, Kantakouzenos opted to go to war with Bulgaria, the ruler of which, John Alexander (1331–1371), demanded that the Byzantines should extradite his predecessor, Michael Šišman. Kantakouzenos replied that Byzantine customs prevented him from satisfying John Alexander's demand and blamed the Bulgarian emperor for initiating a war.[6] It cannot be excluded, however, that the real reason for Kantakouzenos' decision to go to war against the Bulgarians was the opportunity to increase his influence on the army by leading it on a campaign. Moreover, the anticipated war provided Kantakouzenos with the pretext to reform the military *pronoiai*. Apparently, reforms of this kind were the means to secure the support of the soldiers who were maintained through *pronoiai* against his rivals in the competition over the regency of John V, the son of Andronikos III.[7]

Furthermore, the civil war between Andronikos II and Andronikos III, which enabled Kantakouzenos to increase his power and influence, shows that the conflict between the interests of those running the state and those who stood outside the group of immediate family, clan, or individuals who enjoyed imperial favour could lead to explosive situations.[8] It was a war fought between those who possessed the largest shares of the empire's resources under Andronikos II and those who by Michael IX's death lost their hope of gaining imperial favour and their priority in claiming offices and imperial grants. The inse-

---

[5] For the office of *megas domestikos* in the late Byzantine period see S. Kyriakidis "The Role of *megas domestikos* in the Late Byzantine Army," *BSI* 66 (2008), 241–258.

[6] Kantakouzenos, II, 54–55.

[7] Kantakouzenos, II, 58–62.

[8] See Haldon, "Social Élites,"174; For a detailed examination of the causes of the civil war of 1321–1328 and its events see Kyritses, *The Byzantine Aristocracy*, 333–350.

curity, caused by the limited resources of the Byzantine state and its territorial reduction, induced dissatisfied aristocrats and associates of the deceased co-emperor, such as John Kantakouzenos, Theodore Synadenos and Syrgiannes, to rebel actively against the emperor and support the revolt of Michael IX's son, Andronikos.[9]

Although social distinctions in Palaiologan society were clearer than they had been in previous historical periods, the assumption of military command by members of the social elite was not a new development, since the military aristocracy had emerged in the middle Byzantine period. Moreover, it was possible for a limited number of people of low birth to enrich themselves and to climb the social ladder through a successful military career and the limited available evidence suggests that the military elite was not hostile to commanders whose ancestry was not aristocratic.[10] The very negative comments which Kantakouzenos and Akropolites make about Alexios Apokaukos and the Mouzalon brothers reflect only the political circumstances of the moment and personal differences. For instance, Kantakouzenos provides a rather negative image of Apokaukos, who was his main opponent in the civil war of 1341–1347. However, he does not object to the rise of Manuel Tagaris to high office. As Kantakouzenos writes, Tagaris was of low social status, but because he distinguished himself in the constant warfare against the Turks in Philadelphia, he became *megas stratopedarches,* married a niece of the emperor and became a member of the senate, that is to say a leading aristocrat.[11] That bishop Matthew of Ephesos presents a completely different picture of Tagaris' military skills does not affect Kantakouzenos' belief that soldiers of lower social status could advance up the social and military hierarchy.[12] Akropolites was the only one of these new men who survived Michael

---

[9] For Theodore Synadenos see *PLP*, 27125; C. Hannick - G. Schmalzbauer, "Die Synadenoi: Prosopographische Untersuchung zu einer byzantinischen Familie," *JÖB* 25 (1976), 136–137.

[10] For social mobility in late Byzantium see G. Weiss, *Joannes Kantakuzenos—Aristokrat, Staatsmann, Kaiser und Mönch—in der Gesellschaftsentwicklung von Byzanz im 14. Jahrhundert* (Wiesbaden, 1969), 57–60; Matschke-Tinnefeld, *Gesellschaft,* 27–28.

[11] Kantakouzenos, I, 91; Weiss, *Kantakuzenos*, 58. There is no evidence to suggest that in the late Byzantine period there was any attempt to recreate the title and function of senator. The use of the term senate reflects the archaic terminology of the sources but not any institutional change. See the detailed discussion of Kyritses, *Byzantine Aristocracy*, 53–66.

[12] Matthew of Ephessos, *Briefe*, 106, 108, 206.

VIII's coup. Therefore, it was dangerous in his *History* to have inserted positive comments about the Mouzalon brothers. It should not be forgotten that Michael VIII had them murdered in order to ascend the throne.[13]

At the same time as the militarisation of the imperial office came the adoption by the Byzantine elite of an ideal of heroic individualism. This ideal promoted valour on the battlefield, which was displayed through daring acts against the enemy, coupled with excellence in physical activities such as horsemanship and hunting.[14] The lengthy poem which Manuel Philes compiled for John Kantakouzenos is primarily an encomium of the latter's skills in the art of war, and his physical prowess. Praising Kantakouzenos' hunting skills, Philes calls him 'a courteous knight who is chasing his prey' and 'he stifles with the point of his spear everything that is difficult to fight at close quarters.'[15] Shortly afterwards, Kantakouzenos is praised for his deeds against the Turks.

> When an army of barbarians threw Thrace into confusion you teemed with arms and headed to achieve great deeds. You were standing still in the middle of the battle line roaring as a lion ready to attack its prey. And you killed with your sword the invaders.[16]

Michael Doukas Tarchaneiotes Glabas in turn is depicted as a wise general whose sword was painted with blood and whose armies secure victories before the battle begins. He is also praised for slaughtering the Bulgarians as if they were sheep.[17] In a prayer commissioned by Syrgiannes, Philes wrote, 'Boasting of you as my protectors I do not shudder to see the sword even if I am left behind fighting against foreigners you will stand in the first rank as my supporters.'[18] Similarly, the funerary epigram which is still in place above the tomb of the *megas konostaulos* Michael Tornikes insists on his military successes:

---

[13] See Macrides, *The History*, 19–28.

[14] A. Kazhdan, "The Aristocracy and the Imperial Ideal," in M. Angold (ed.), *The Byzantine Aristocracy IX to XII Centuries* (Oxford, 1984), 52.

[15] Philes, I, 144.

[16] Philes, I, 149.

[17] Philes, II, 56–57, 248. For Michael Doukas Tarchaneiotes Glabas see *PLP*, 27504; Polemis, *The Doukai*, 121.

[18] Philes, I, 244–245.

> And upon enemies he breathed a strategic flame (such as military rules required) and was an irresistible thunderbolt upon their serried ranks. He presided over the army like a father guarding the commonweal lest any advantage to it should be stolen. Contracting a high-born and seemly marriage connection, and securing thus royal affinity.[19]

The cult of the martial virtues did not prevent Byzantine military leaders from seeking distinctions in the sphere of letters. The *megas domestikos*, emperor, monk and historiographer John Kantakouzenos is a well-known case. Wishing to promote Kantakouzenos' military skills, Philes depicts him as a great warrior, an 'invincible giant in battles.' Shortly afterwards he praises him for compiling 'works of wisdom.' Michael Tarchaneiotes Glabas is said to have compiled theological works. He was the patron of the Pammakaristos monastery and after capturing Sozopolis from the Bulgarians he founded the monastery of St John.[20]

Beyond the poems of Philes, examples praising individual bravery on the battlefield can be found in historical accounts which show great men turning the tide of battle by their prowess and by cutting a path through their enemies. Akropolites' and Gregoras' descriptions of the fighting between Theodore I and the Seljuk sultan in the battle of Antioch-on-the-Meander extol individual heroism on the battlefield. During the course of the battle, 'being confident to the strength of his body' the sultan was seeking to fight the emperor in person. When the sultan approached, Laskaris inflicted on him a powerful blow. Then Laskaris fell from his horse. Theodore I managed to recover rather quickly. While the sultan was ordering his soldiers to capture him, 'Theodore, who had fallen from his horse as dead, suddenly stood up and, filled with rage and enthusiastic impulse,' he took out his sword and hit the sultan's horse. The sultan fell from his large horse 'as if he was falling from a tower.' He was immediately beheaded by the emperor, according to Gregoras' and Choniates' sources; by an unknown soldier, according to Akropolites' source.[21] Describing the battle of Rosokastron, where in 1332 the Bulgarians under the leadership of their ruler John Alexander inflicted a heavy defeat on the

[19] A.Van Millingen, *Byzantine Churches in Constantinople* (London, 1912), 330–333, whose translation this is. For Michael Tornikes see *PLP*, 29132.

[20] Philes, I, 180, II, 230–233, 245.

[21] Akropolites, I, 16–17; Gregoras, I, 20–21; Choniates, *Orationes*, 171.

Byzantines, Gregoras comments that Kantakouzenos sustained many blows inflicted on him by the swords and spears of the enemies. Yet he managed to overcome many enemies. Throughout the battle, he lost neither his shield nor his sword. He still remained firmly on his horse. Gregoras adds that the second bravest warrior after Kantakouzenos was the *protosevastos*, son of the caesar (John Palaiologos) and grandson of the *porphyrogennetos* (Constantine Palaiologos, son of Michael VIII). Gregoras comments that although his horse was seriously wounded after receiving many blows he did not bear the shame of retreat.[22]

These accounts of heroic deeds on the battlefield are similar to descriptions of military conflict in the wider geographical context of the Byzantine world. Recording the deeds of Leonardo Tocco, the anonymous author of the *Chronicle of the Toccos*, who was a proud 'Roman,' relates how in 1402 his hero sustained the blows inflicted on him by nine lances and his horse was killed after it was repeatedly hit by the lances of his Albanian enemies in Angelokastro. However, brandishing his sword and fighting against many enemies, Tocco managed to cut through the lines of the enemy and save his life. In another instance, Tocco is portrayed as standing on the saddle of his Calabrian horse like a sun. As the anonymous author remarks, he looked like a tower among his troops.[23]

The ideal of heroic individualism is intensively promoted by the late Byzantine romances, which are not merely works of entertainment. They mirror ideals which were amongst society's ideals in this period and their content reflects the values and culture of the higher aristocracy.[24] The *Achilleid* is a work that can be connected to the court of Neopatras of the late thirteenth and early fourteenth centuries and therefore, it would appear to be precluded in Byzantine territory adjoining Frankish Greece.[25] Its hero is not only an excellent general and clever tactician. On the battlefield he displays great bravery, which often results in the mass killing of the enemies. Contrary to the sug-

---

[22] Gregoras, I, 486–487. He was the son of John Palaiologos and Eirene Metochites, daughter of the *megas logothetes*, Theodore: Sevčenko, *Études*, 149.

[23] *Cronaca dei Tocco di Cefalonia di Anonimo,* G. Schiró (ed.), (Rome, 1975), vv. 340–348.

[24] D. Jacoby, "Knightly Values and Class Consciousness in the Crusader States of the Eastern Mediterranean," *MHR* 1 (1986), 166.

[25] *The Byzantine Achilleid. The Naples Version*, O.L. Smith (ed.), (Vienna, 1999), vv. 590–602, 620–625; Magdalino, *Between Romaniae*, 89.

gestions of his brothers, Kallimachos, who is called invincible in battle and great in war, decided to continue his march through an inhospitable mountain. He wonders how he would stand in a battle and achieve heroic deeds if a mountain can defeat him. He cannot accept that he might not inherit the empire because he could not overcome such difficulties. In this case, their father should not bequeath the empire to any of his sons and relatives, but to a stranger who is a brave soldier. In addition, the hero feels shame before his generals and soldiers. Eventually, it is decided to let most of the army and the baggage train go and to resume their march, carrying only what is necessary.[26] The heroes of the *War of Troy*, like Iasou, are called brave and stronghanded soldiers who have fought many fierce battles from which they have emerged victorious. The accounts of battles focus on the deeds of individual heroes.[27]

The role of Western European military ideals and practices

Was the development of a military ethos which emphasised valour in combat influenced by Western European military ideas and ideals? Although military manuals exhort the general not to participate in the fighting himself because a disaster to him might cause the whole army to be demoralised, long before 1204, Byzantine authors were not reticent about praising the heroic acts of emperors and generals.[28] As an emperor and aristocrat, Kantakouzenos, throughout his *History*, promotes ideas about war which are quite similar to Western European ideas. Honour (*time*) is the main one. During the conflict with Bulgaria he stated that it is an honour for the Byzantines not to be despised by the 'barbarians.' Kantakouzenos also expressed the wish not to be

---

[26] *Kallimachos and Chrysorrhoe*, in E. Kriaras (ed.), *Τα μεσαιωνικά ιπποτικά μυθιστορήματα* [*The Medieval Chivalric Romances*] (Athens, 1955), vv. 101–145. *Kallimachos and Chrysorrhoe* is attributed to Michael VIII's nephew, Andronikos Komnenos Doukas Angelos Palaiologos: R. Beaton, *The Medieval Greek Romance* (Cambridge, 1986), 101.

[27] *The War of Troy*, M. Papathomoupoulou - E. Jeffreys (eds.), (Athens, 1996), vv. 57–59, 940–1029.

[28] G.T. Dennis, "The Byzantines in Battle," in N. Oikonomides (ed.), *Byzantium at War (9th–12th Centuries)* (Athens, 1997), 174; J. Haldon, "Blood and Ink: Some Observations on Byzantine Attitudes towards Warfare and Diplomacy," in. J. Shepard - S. Franklin (eds.), *Byzantine Diplomacy* (London, 1994), 287.

outstripped by anyone in his offering of servants, money, and of himself 'for the honour of the Byzantine emperors.'[29] In another example, describing the games organised by Andronikos III on the occasion of his marriage, Kantakouzenos remarks that the participants sought honour and he is proud to say that the emperor has proved better than his Western European opponents.[30] He also contrasts fighting for honour with fighting for money. Kantakouzenos told the Serbian ruler that for the Byzantines it is dishonourable to serve for pay and not for the love of honour and friendship.[31] Similarly, he promotes the value of defending the fatherland and the Christians, considering it his duty. He states that one of the main duties of the emperor is to thwart the 'Ismaelites.' He also comments that he wished to see the Turks being punished for the sufferings that they inflicted on the Christians.[32] The ideals described in the *History* of Kantakouzenos do not differ substantially from those praised by the prince of Achaia, William II of Villehardouin, as they are stated in the *Chronicle of the Morea.* According to the anonymous author on the eve of the battle of Pelagonia, William II commented that his aim was to augment his honour, wealth and fame. To achieve this he should not be unjust and grab from his relatives and friends. The prince concludes that he should be praised for doing so because these are the ideals that every soldier ought to seek.[33]

Furthermore, both the Byzantine and Western elites had the same goal, the maintenance of their privileges and status; the possession of high military office was a means of achieving this. For this reason, the Byzantine military aristocracy did not hesitate to accept some Western European practices related mainly to the display of superiority and class exclusiveness. It is logical to conclude that the adoption of chivalric practices was useful not merely for the display of power. It also influenced and corresponded to the military ethos of at least a significant part of the late Byzantine aristocracy. In this context, the Byzantine ruling elite promoted activities such as tournaments and jousting. Both tournaments and jousting have their origins in eleventh-century warfare, and apart from martial prowess and the promotion

[29] Kantakouzenos, I, 184–185.
[30] Kantakouzenos, I, 204–205.
[31] Kantakouzenos, III, 143.
[32] Kantakouzenos, III, 59.
[33] *Chronicle of the Morea*, vv. 4141–4156.

of class exclusiveness, they were used as part of the training that the soldiers were expected to receive.[34]

With reference to late Byzantium, it has been argued that,

> The brief appearances of sport of this kind in a culture, where ideals of chivalry were known, but regarded an alien concept, indicate the furthest and most unexpected limit of the tournament's appeal.[35]

However, war games were no innovation in late Byzantium. Long before a Byzantine emperor participated for the first time in a tournament, Nikephoros II Phokas held in 966 a cavalry contest (*hippikos agon*) in the hippodrome of Constantinople which terrified the townspeople, who, according to Leo the Deacon were ignorant of military activities. Leo also comments that Nikephoros' aim was to train his troops for war.[36] Undeniably, the historical, social and political context in which Nikephoros II organised this event was very different from the period of the tournaments. Moreover, this was a show in which the participants were regular soldiers and not exclusively members of the aristocracy, who wished to promote their sense of class exclusiveness. Nonetheless, Nikephoros II was a member of a leading family of tenth-century military aristocrats and events such as the one mentioned above, which was then far away from the battlefields of the frontier zones, were aimed at the glorification of martial skills, which the military aristocracy of the tenth century promoted and used as its proving ground. Therefore, since one of the aims of the tournament and jousting was to promote and glorify the military skills of the ruling elite, there are some similarities between Nikephoros II's display and the tournaments.

Since Manuel I Komnenos (1143–1180) introduced Western European military practices into Byzantium, it is not surprising that he was in 1159 the first Byzantine ruler to participate in a tournament.[37]

---

[34] M. Keen, *Chivalry* (Yale, 1984), 190; J.F. Verbrugen, *The Art of Warfare in Western Europe during the Middle Ages*, tr. S. Willard - S.C.M. Southern (Suffolk, 1997), 30–36.

[35] R. Barber - J. Barker, *Tournaments, Jousts, Chivalry and Pageants in Middle Ages* (Woodbridge, 1989), 105.

[36] Leo the Deacon, *Leonis diaconis Caloensis Historiae libri decem*, C.B. Hase (ed.), (Bonn, 1828) 63; A.M. Talbot - D.F. Sullivan, *The History of Leo the Deacon* (Washington, 2005), 46–47, 112.

[37] Choniates, *Historia*, 108. Choniates states that when Manuel I entered Antioch in 1159 he participated in a tournament with the local crusader knights. He was head of the Byzantine contestants, who were all his relatives. An anonymous text provides a

In late Byzantium there is evidence that such activities were organised. Moreover, the establishment in Frankish Greece of Western knights, such as William Bouchart, who according to the French version of the *Chronicle of the Morea*, was one of the best jousters in the West, made the Byzantines more familiar with jousts and tournaments.[38] Two of them were held by Andronikos III, as part of the celebration of his wedding and the birth of his heir apparent. They were probably small-scale events, apparently without any great military or political significance. Kantakouzenos noted that the participants sought honour; he does not say anything about military training.[39] Perhaps it was not a coincidence that both events were held at a time when tournament and jousting were widespread in Italy and in Savoy, the home of Andronikos III's wife.[40] In 1326, Anna of Savoy at her marriage to Andronikos III was accompanied to Constantinople by mounted esquires and nobles; they, as Kantakouzenos states,

> Taught jousting and tournament to the Romans, who were ignorant of these games. Therefore, many Romans practiced these games seeking honour and particularly the emperor, who proved himself superior to his tutors. So, the Savoyans, French, Alamans and Burgundians conceded defeat.[41]

Kantakouzenos' description of this event, his statement that the participants sought honour, that Andronikos III proved superior to them, as well as the fact that he mentions the specific geographical origins of the Westerners, (the teams participating in the games were always formed on the basis of their geographical origins) show that not only did he approve the holding of this event, but also he fully understood its social significance. Kantakouzenos, Andronikos III and logically other people of their circle seem to have believed that tournament and *joustra* corresponded to the social structures of fourteenth-century Byzantium and to the military character of the late Byzantine aristocracy. Moreover, there is no evidence to indicate the existence

---

detail description of the garments worn by Manuel I in jousts: H. Maguire – L. Jones, "A Desccription of the Jousts of Manuel I Komnenos," *BMGS* 26 (2002), 104–148.

[38] Jousts and tournaments were held by the Frankish knights of the principalities of Achaia and Athens: *Chronicle of the Morea*, vv. 2408, 3368–3370; *Livre de la conqueste*, ch. 1016–1017 See also the discussion of Jacoby, "Knightly Values," 158–186.

[39] Kantakouzenos, I, 204–205.

[40] Barber-Barker (*Tournaments, Jousts, Chivalry*, 79–80) state that between 1320 and 1340 there are references to eight tournaments in northern Italy.

[41] Kantakouzenos, I, 204–205.

in Byzantium of tournaments held by wealthy aristocrats without the consent of the emperor. Instead, the participants were the emperor, his relatives and close associates.

In contrast, Gregoras demonstrates the dichotomy between the traditional Byzantine ideological system and such sporting activities. Describing the games that Andronikos III organised in 1331, he remarks that 'they had been invented by the Latins long ago, to exercise their bodies when they were not involved in war.' Gregoras states that the emperor many times nearly escaped serious injury and was advised, mainly by the older people, to abstain from such games; he was told it was not proper for an emperor to be involved in games with servants. Before giving a detailed and precise description of the tournament, Gregoras says that Andronikos III considered the admonitions of the older people sheer cowardice.[42] It seems that Gregoras and other Byzantines found Andronikos III's participation in such activities inappropriate for the imperial prestige. However, Gregoras does not speak against the idea of holding these games. What he opposes is imperial participation in them. It should not be forgotten that Gregoras disapproves of the manner of Andronikos III's ruling; he accuses Andronikos of neglecting old imperial customs, which are threatened with extinction and of indifference to the protocol of official ceremonies.[43]

Andronikos III was not the only late Byzantine emperor who organised tournaments. There is additional evidence for tournaments in the account of the Burgundian traveler, Bertrandon de la Brocquière. He writes that in 1432 as part of the celebrations following the marriage of a member of the imperial family, he attended a tournament, which, however, was different from those held in his native land. He comments that the participants did not bear armour, but used sticks instead of swords and that no one was injured. This example shows that we cannot exclude the possibility that tournaments, even in a different form from the Western European ones, became common in late Byzantium. It is particularly interesting that, apart from the tournament mentioned above, Bertrandon de la Brocquière reports that he saw in the hippodrome the despot of the Morea, the emperor's brother, practicing together with other forty mounted archers in the

[42] Gregoras, I, 482.
[43] Gregoras, I, 565–567.

fashion of the Turks. More specifically, each horseman threw his hat before him and shot at it as soon as he had passed it. Bertrandon concludes that the one whose arrow pierced his hat was esteemed the most expert.[44]

Beyond the above references there is no further available information concerning the holding of jousting and tournaments in late Byzantium, although famous enthusiasts of such activities and of chivalric idealism, like Amadeo of Savoy and Jean le Meingre Boucicaut, came to Constantinople in 1366 and 1399 respectively with their own troops and fought in Byzantium's service.[45] Moreover, it should not be surprising that all the known emperors who have participated in war games, Nikephoros II Phokas, Manuel I Komnenos and Andronikos III Palaiologos, were proud soldiers with significant military experience. Tournaments and jousts provided members of the social elite with the opportunity to promote their martial prowess and enrich the festivities that followed great occasions. However, unlike Western Europe, late Byzantium had no individual warriors who sought enrichment through the rewards and ransoms involved in these games.

Although martial sports were attributable to Western influence, sports were not unknown in Byzantium; the *tzykanion* (ball game) was the sport of the ruling elite. Therefore, it promoted like *joustra* and tournaments the class exclusiveness of the higher aristocracy, but not its military character. It is worth noting, that although there is no evidence that the Nicaean rulers ever participated in or organised such activities, writing in the 1250s Nikephoros Blemmydes is very critical of them. While he recommends that the ruler should intensively train in war skills, he doubts whether duels and games such as the *tzykaton*, or war games of any kind are of any practical use.[46] However, almost a century later Kantakouzenos is praised by Manuel Philes for his performance in sports, mainly in *tzykanion*.[47]

---

[44] Bertrandon de la Brocquière, *La voyage d'Outremer de Bertrandon de la Brocquière*, Ch. Schefer (ed.), (Paris, 1892), 158–166.

[45] For the achievements of Amadeo and Boucicaut in tournament and jousting and for their ideas about chivalry see Ph. Contamine, "Les tournois en France à la fin du moyen âge," in J. Fleckenstein (ed.), *Das ritterliche Turnier im Mittelalter* (Goettingen, 1985), 438; Barber-Barker, "Tournaments, Jousts, Chivalry," 80; N. Housley, "One Man and His Wars: The Depiction of Warfare by Marshall Boucicaut's Biographer," *JMH* 29 (2003), 27–40.

[46] *PG* 142, cols. 636–637.

[47] Philes, I, 150, 180–181.

Beyond Gregoras' criticism and Blemmydes' objections, nothing is known about the reactions that these activities might have caused in late Byzantium. For instance, in the West the church was hostile towards these activities, and since 1130 had repeatedly issued orders banning them, although these bans were never enforced. Eventually, they were revoked in 1316, a few years before the known late Byzantine tournaments took place.[48] No surviving information suggests that the Byzantine church during the period under discussion officially opposed war games, unless Gregoras' 'older people,' who opposed Andronikos III's participation, were members of the higher clergy. The lack of any opposition from the Byzantine church, like the ineffectual Western European bans, may strengthen the view that *joustra* and tournaments were not widespread in late Byzantium, or, more probably it reflects the scarcity of sources. Any conclusion entails much speculation.

Modern scholars debate whether or not heraldry existed in Byzantium. Heraldry has been defined as 'the systematic use of hereditary insignia on the shields of noblemen, which started following established rules from the 12th century.' Later, heraldic devices were displayed on the surcoats of knights, on the trappings of their horses, on their seals and on their tombs and effigies.[49] A study published in the 1920s concludes that no Byzantine family used heraldic emblems.[50] Nevertheless, the presence of Western European knights in Frankish Greece and the existence of Latin mercenaries in the Byzantine army, who were admitted to the Byzantine nobility, must have made late Byzantine aristocrats familiar with Western European heraldry. Moreover, the cultural encounters in late medieval Mediterranean between Westerners, Byzantines and Muslims resulted in the development of what has been defined as a common 'language of power' among the Mediterranean elite.[51] This meant that social groups of these diverse civilisations adopted signs and insignia which connoted their high status and prestige in a language that could cross cultural and

[48] D. Carlson, "Church Councils on Chivalry," in. H. Chikering – T.H. Seiler (eds.), *The Study of Chivalry. Resources and Approaches* (Kalamazoo, 1988), 144; Contamine, "Tournois en France,"427; Barber-Barker, *Tournaments, Jousts, Chivalry,* 139–145; R. Kaueper, *Chivalry and Violence in Medieval Europe* (Oxford, 1999), 80.

[49] Keen, *Chivalry*, 125–126.

[50] G. Typaldos, "Εἶχον οἱ Βυζαντινοί οἰκόσημα;" [Did the Byzantines have Coats of Arms?] *Epeteris Etairias Vyzantinon Spoudon* 3 (1926), 206–222.

[51] R. Ousterhout, "Byzantium between East and West and the Origins of Heraldry," in C. Hurihane (ed.), *Byzantine Art. Recent Studies* (Tempe, 2009), 170.

linguistic boundaries.[52] In their investigation of the mosaics and frescoes of the Pammakaristos monastery (Fethiye Camii) in Istanbul, Mango, Belting and Mouriki conclude that the recurrence of the rampant lion in round medallions indicates that it is a family crest.[53] It is also interesting that the founder of the monastery was Michael Doukas Glabas Tarchaneiotes, who was a renowned general. The image of the rampant lion was rather common in symbolic imagery across Western Europe and the Mediterranean.[54] Moreover, a crowned rampant lion bearing a sword with a monograph of a Palaiologan ruler once was visible in the sea walls of Constantinople, while the *Tzakones*, who were a palace guard unit, wore distinctive breastplate with lions.[55] Furthermore, in their investigation of the monastery of Lips (Fenari Isa Camii), Mango and Hawkins suggest that certain simple linear devices which accompany the monographs of the families of the Palaiologoi, the Asan and the Kantakouzenoi and are found on sarcophagi, coins and architectural fragments prove the existence of heraldic emblems in Byzantium.[56] It could be argued that their simplicity implies that most probably they were ornamental devices and not heraldic emblems.[57]

While the aforementioned findings have been identified as heraldic symbols, there is no evidence that the rampant lion and other emblems, which were similar to the heraldic emblems used in the West, were bestowed by the emperor or codified according to a person or family.[58] Moreover, we do not know whether these emblems were transmitted among descendants like Western European heraldic symbols. Furthermore, there is no evidence to suggest that the arrangement and description of heraldic devices in Byzantium was regulated by well-defined rules. Nor do we know whether Byzantine soldiers and military leaders used heraldic emblems that recalled their heroic deeds and promoted their courage and bravery on the battlefield. It is worth noting that in the course of the thirteenth century heraldry stopped

[52] See S. Redford, "A Grammar of Rum Seljuk Ornament," *Mesogeios* 25/26 (2005), 283–310; Ousterhout, "Origins of Heraldry," 169.

[53] H. Belting – C. Mango – D. Mouriki, *The Mosaics and Frescoes of St Mary of Pammakaristos (Fethiye Camii) at Istanbul* (Washington, 1978), 21–22.

[54] Ousterhout, "Origins of Heraldry," 157–158.

[55] A. Van Millingen, *Byzantine Constantinople* (London, 1899), 189; Pseudo-Kodinos, 181.

[56] C. Mango – E.J.W. Hawkins, "Additional Finds at Fenari Isa Camii," *DOP 22* (1968), 181.

[57] Kyritses, *Byzantine Aristocracy*, 248.

[58] Ousterhout, "Origins of Heraldry," 159.

being the preserve of the greater aristocracy and became emblematic of the pride of birth and culture of the nobility in its broader range.[59] However, the Byzantine archaeological findings, which have been identified as heraldic emblems, are associated with members of the higher aristocracy.

As far as the narrative sources are concerned, in his description of the preparations of the Byzantine campaign against Martino Zaccaria, the Genoese ruler of Chios, Kantakouzenos may be implying the use of coats of arms by the Byzantines. He remarks that the soldiers decorated their shields and arms with their insignia (*parasema*) because, as he writes, in this campaign participated not only the common soldiers but also the noble and very powerful.[60] It is likely that these insignia were the family emblems of high aristocrats, indicating their pride of birth, high status and martial achievements. Furthermore, if we believe a Western source, Kantakouzenos was aware of the use and meaning of heraldry in the West. According to a French chronicle, in 1347, John Kantakouzenos permitted the mercenaries William Poujoize and John Bruidy, nobles of the city of Metz, to change the swallows in their arms into eagles, in recognition of their good service against the 'Saracen.'[61] Moreover, it is likely that the ceremony of conferring the title of *kavallarios* upon Western European mercenaries involved the granting of the right to use armorial devices which promoted the high status, sense of honour and military achievements of the *kavallarioi*.

While the extent to which the Byzantines adopted heraldic emblems remains unknown, the later Byzantine aristocrats extensively used monograms of family names as an indication of pride of birth and connection to the ruling elite. It has been concluded that the extensive use of monographs shows that the emblems of Byzantine distinctiveness were based on written word and not on pictorial images.[62] The monograms of family names were an indication of pride of birth and connection to the ruling elite. According to Oikonomides, sigilliographic evidence suggests that the use of monograms was common

[59] Keen, *Chivalry*, 127–128.

[60] Kantakouzenos, I, 375.

[61] *Les chroniques de la ville de Metz*, J. Huguenin (ed.), (Metz, 1835), 83; J. Schneider, "Sir Nicole Louve: citain du Metz," in P. Contamine (ed.), *La noblesse au moyen âge* (Paris, 1976), 183.

[62] Ousterhout, "Origins of Heraldry," 165.

between the sixth and eight centuries. They became common again under the Komnenoi and mirror the growth of a new aristocracy in which blood ties with the imperial or other aristocratic families was of paramount importance.[63] A good example of a monogram bearing the name of an elite aristocrat of the later Byzantine period is provided by a seal of Theodore Synadenos.[64] A characteristic development of the later period was the use of monograms on the very clothes worn by the aristocrats. Such examples can be found in the portraits standing over the tombs in Kariye Cami. In the garment of Eirene Raoulaina Palaiologina, there are three distinct monograms. The most frequent ones are the monographs of the Palaiologoi and of the Asan-Palaiologoi, who were the descendants of Andronikos II's sister Eirene and her husband John III Asan of Bulgaria in 1278. These monograms appear also in two other tombs. The third monogram is that of the family of Raoul.[65] The funeral portraits of the frescoes of Kariye Cami show that more than one monogram per person could be used to promote and clarify a relationship by blood or marriage to the ruling elite.

The effectiveness of Byzantine generals on the battlefield

To evaluate the effectiveness of the late Byzantine generals, certain constraints should be borne in mind. For most of them, battlefield experience was limited. Very few military commanders had led troops in a significant number of battles or military operations of any kind for longer than a decade, as John Palaiologos and Michael Doukas Glabas Tarchaneiotes had done. This development was no different from what was happening in Western Europe, where military experience, even for a king, was limited.[66] Moreover, almost all military commanders were chosen from a corps of officers who were recruited primarily not for their valour and military skills but because they conformed to specific social criteria. This inevitably caused a conservative attitude among

---

[63] N. Oikonomides, *A Collection of Dated Byzantine Lead Seals* (Washington, 1986), 153–154; idem, *Byzantine Lead Seals* (Washington, 1985), 23.

[64] W. Seibt, "Das Monogram-Siegel eines Theodoros Dukas Synadenos aus der frühen Palaiologenzeit," *JÖB* 40 (1970), 272.

[65] Underwood, *The Kariye Djami*, I, 284–286.

[66] J. France, *Western Warfare in the Age of the Crusades. 1000–1300* (London, 1999), 140.

the late Byzantine leaders, since they did not need to improve their skills or to achieve important military successes to rise in the military hierarchy. This situation made it more difficult for the military leaders to learn from the experience of the past. In addition, the lack of sufficient resources in terms of manpower and finance did not allow the Byzantine army to keep up with the administrative, organisational and technological developments of the fourteenth and fifteenth centuries in Western Europe and in the early Ottoman state.

With regard to the training of the late Byzantine commanders, it should be noted that the significance of training was emphasised by the military treatises of earlier times. In addition to recognising training as essential for the maintenance of a reliable military force, they refer to the training of the entire army as a single unit and stipulate that the military commanders should be in charge of the training.[67] The question is how far these ideas on military training apply after 1204. Nikephoros Blemmydes, who was not a military man, in his *Imperial Statue*, emphasises the need for the army to be trained in land and sea warfare alike, and states that the emperor himself should be in charge of the training of his troops.[68] In contrast, Kantakouzenos indicates that training for the aristocratic commanders was personal. He states that he had been taught the art of war, together with his close friend Syrgiannes, by their common uncle Angelos, who was a military man.[69] Kantakouzenos also states that he had taught his nephew John Angelos the art of war, without explaining, what this teaching involved.[70] Manuel II comments that among the compulsory things that he was taught were the handling of the bow and the spear and how to ride.[71] It is certain that Michael VIII Palaiologos and his

---

[67] For references to military training in Byzantine military manuals see *Taktika*, VII. 2; G.T. Dennis, *Three Byzantine Military Treatises* (Washington, 1985), 318; E. McGeer, *Sowing the Dragon's Teeth: Byzantine Warfare in the Tenth Century* (Washington, 1995), 88. For a brief evaluation of military training in the middle Byzantine period see J. Haldon, "Approaches to an Alternative Military History of the Period ca. 1025–1071," in A. Avramea – A. Laiou – E. Chryssos (eds.), *The Empire in Crisis (?) Byzantium in the 11th Century (1025–1081)* (Athens, 2003), 54–55.

[68] *PG* 142, cols. 637–639.

[69] Kantakouzenos, I, 333–334; Nicol identifies Kantakouzenos' uncle with John Angelos Synadenos, the father of the *protostrator* Theodore Synadenos: D. Nicol, *The Reluctant Emperor. A Biography of John Cantacuzene, Byzantine Emperor and Monk, c. 1295–1383* (Cambridge, 1996), 17.

[70] Kantakouzenos, I, 274.

[71] Manuel II, *Funeral Oration*, 104–105; Barker, *Manuel II*, 529; Dennis, *The Reign of Manuel II*, 14.

brothers John and Constantine were been taught much by their father, Andronikos, who was a leading commander of John III. Alexios Raoul campaigned against the Turks in Asia Minor, together his brother and his son, John.[72] Consequently, it seems safe to conclude that the custom for all the military commanders of the aristocracy was that they were trained in fighting and commanding troops by their close relatives. Almost all of them had an ancestor who had been a commanding officer. The personal training of the elite does not seem to have been a new development in late Byzantium. Members of the families of the military aristocracy in the middle Byzantine period were probably trained by their relatives. For instance, it is logical that the emperor Nikephoros II Phokas learnt a great deal from his father, the general Bardas Phokas.

Furthermore, describing the skills and qualifications of military leaders, late Byzantine sources refer to their training and experience more through their long involvement in military operations and less through practice and exercise outside war. Success was seen as a result of experience and not of theoretical knowledge. Pachymeres states that in 1281, Michael VIII, wishing to train his son Constantine in warfare sent him to lead a campaign against the Serbians in Macedonia.[73] An imperial panegyric praises the same emperor for his skills in handling the spear and the bow, which he acquired not through teaching but long practice.[74] Other examples refer to the experience of military commanders; Akropolites states that Nikolaos Kotertzes, the general sent by John III to defend Tzouroulou against the Latins had been tested in many wars and was so successful that everyone conceded that no one before him had managed to achieve so much.[75] The governor of Nicaea in 1263, Nikolaos Manuelites, is praised as an experienced soldier.[76] On the eve of the campaign in Bithynia, in 1329, Andronikos III asked the advice of the governor of Mesothynia, Kontophre, who, according to Kantakouzenos, had experience of many wars, in particular those against the Turks.[77]

[72] Philes, I, 440–441. For Alexios Raoul see *PLP*, 24115.
[73] Pachymeres, II, 599.
[74] *AG*, I, 326–327.
[75] Akropolites, I, 51.
[76] Pachymeres, I, 323.
[77] Kantakouzenos, I, 341.

Does the fact that personal experience was the primary source of knowledge mean that military manuals were unknown? It has been stated that one of the difficulties with which the Western military commander had to deal was that he was not served by an institutional memory and contemporaneous Western European authors comment that generals were taught the art of war through practice and not through any form of teaching.[78] However, late Byzantine generals had at their disposal military treatises, which, although written centuries before our period, would have been known and available.[79] The manuscript tradition of Byzantine military treatises, in particular of Leo's *Taktika*, is strong evidence that these works were in circulation in late Byzantium.[80] It is logical that late Byzantine generals read a significant part of the large number of manuscripts containing military treatises which circulated among them. Michael Doukas Glabas Tarchaneiotes, a general with long military service in the late Byzantine period, is said to have written a military treatise, which has been lost.[81] It would be interesting to know whether Glabas Tarchaneiotes' treatise was a repetition and imitation of older manuals, or reflected contemporary military developments and the personal experiences of the author, such as fighting against French heavy cavalry, which he encountered in Epiros and Albania, and technological advances, such as the increasing use of the crossbow. However, the lack of sufficient sources and the absence of late Byzantine military treatises make it difficult to understand how later Byzantine generals treated the military manuals compiled by their ancestors. It is reasonable to conclude that the technological developments and the changes that occurred in the art of war from the thirteenth through the fifteenth centuries would have made the practical advice of the military manuals of earlier epochs obsolete. Nonetheless, these works are likely to have attracted the literary interest of highly educated military commanders, such as Michael Doukas Glabas Tarchaneiotes, John Kantakouzenos and Manuel II. Late Byzantine

[78] France, *Western Warfare*, 186; Settia, *pratica e teoria*, 20–26, with a discussion of the sources.

[79] See Kekaumenos, *Cecaumenos strategikon*, B. Wassiliewsky – V. Jernstedt (eds.), (St. Peterburg, 1896 repr. Amsterdam, 1965), 19. Although he is writing in the eleventh century, long before the period under discussion, he suggests that the general should read military and history books when he has leisure time and is not occupied by any duties.

[80] See A. Dain – J. de Foucault, "Les stratégistes Byzantins," *TM* 2 (1967), 317–392.

[81] See Polemis, *The Doukai*, 121.

authors, Kantakouzenos, in particular, an experienced soldier, do not make direct references to older military manuals. Nevertheless, as will be discussed below, some principles outlined by the military treatises of earlier epochs are echoed in late Byzantine authors. This cannot be seen simply as a repetition of the instructions prescribed in the older military manuals. It rather reflects the fact that the late Byzantine generals faced similar challenges to their ancestors, such as attacks from mounted bowmen.

Furthermore, in spite of being written in different political contexts, the military treatises and historical accounts of earlier periods in Byzantine history recognised the imbalance of resources between Byzantium and its enemies. Generals are exhorted to avoid fighting in unfavourable conditions and to avoid wasting manpower and resources. Instead, armies should proceed with the utmost caution; generals preferred to use craft and intelligence.[82] Similarly, aware of the unviable strategic position of their state, late Byzantine authors exhort the generals to do their best to avoid unnecessary risks, which would result in the waste of valuable resources. Victory could be achieved through a combination of delaying tactics, stratagems and the exploitation of enemy weaknesses, while generals should do their best to protect their manpower. Kantakouzenos remarks that the good general is not only the one who achieves victories but also the one who when he fails does not lose his forces.[83] To give another instance, the same author states that the most prudent of the emperors did not always use bravery to encounter the enemies in pitched battle; they also used deception, cunning methods and stratagems, in particular when they are engaged with fearful and very strong enemies; no one reproaches them for cheating. On the contrary, everyone admires those methods which achieve more than bodily strength could.[84] Thomas Magistros exhorts the emperor to keep the aim of his military operations secret. He suggests that when the emperor campaigns against the Persians (Turks) he should pretend that he is marching on the Ethiopians (Mameluks), and if he wishes to fight the Bulgarians, he should make preparations to march on the Serbians.[85] Kantakouzenos did exactly this in 1341. His intention was to march on Adrianople. He

---

[82] Haldon, "Blood and Ink," 281–294; Dennis, "The Byzantines in Battle," 165.
[83] Kantakouzenos, I, 103.
[84] Kantakouzenos, II, 116.
[85] *PG* 145, col. 466.

did not want his plans to be revealed and therefore, pretended that he was marching on other Thracian cities.[86] In the Funeral Oration to his brother, the despot of Morea, Theodore I, Manuel II, writes,

> Need I say that it is a law of war that a man putting his enemy to flight by deceit should be praised more than if he were to defeat him in open battle? For in the first case the general wins a victory without any losses, while in the second case he is compelled to sacrifice many of his men.[87]

Although it is highly likely that individuals such as Kantakouzenos and Manuel II read old military manuals, these statements do not suggest the direct influence of these military treatises. Rather they reflect similarities in the military and political context and the continuation in the later period of traditional Byzantine attitudes towards warfare, which emphasised the avoidance of military conflict. Moreover, precepts such as avoidance of battle, passive resistance, complemented by harassment tactics aimed at leading the enemy to overextend his lines, attrition of the enemy by trying to deprive him of water and forage, and the use of spies and deserters for the dissemination of false rumours, appear in the Hellenistic and Roman treatises.[88]

The planning of the Byzantine tactics at the battle of Pelagonia (1259) exemplifies the above principles. The Byzantines employed delay and guerrilla tactics, damaging the supplies, coherence and morale of the enemy, mainly of the Western European knights, whose strength was based on discipline. Therefore, the outcome of the battle was decided by good leadership. The Byzantine military command managed through the proper tactical formation to reverse the expected outcome of the battle. The Byzantines, taking advantage of the geography, gathered the right intelligence concerning the strength, size, weapons and also weak points of the enemy. Moreover, the employment of such tactics demanded a high degree of discipline, which the Byzantine generals managed to impose. The battle of Pelagonia proves that good leadership may overcome technological inferiority. The Latin knights bore armour and weapons that were much more expensive and heavier than

[86] Kantakouzenos, II, 187.

[87] Manuel II, *Funeral Oration*, 189.

[88] J. Haldon, *Warfare State and Society in the Byzantine World. 565–1204* (London, 1999), 37–38; Hunger, *Literatur*, II, 324–338.

those borne by the cavalry and foot archers of the Byzantine army.[89] In another example, on the eve of the civil war, in 1321, Kantakouzenos sees as a prudent act the refusal of the general Manuel Tagaris to obey Andronikos II's order to attack the troops of Andronikos III in Adrianople, because he did not know the strength of the enemy.[90]

It is particularly interesting that ideas about prudent generalship can be found in romances, which reflect the class values and military ethos of the higher aristocracy. Although their authors promoted heroic deeds on the battlefield, they admit that such acts of bravery are not in agreement with prudent military thinking. For instance, Achilleus rejects his general's advice to take the enemy by surprise at night. He argues that while this is good advice and correct in military terms, a night attack is not the act of brave and strong soldiers. He claims that even if they are victorious, such an action is shameful.[91] Kallimachos' brothers hesitate to attack the heavily fortified castle, because to fight when there is no need to do so is something that the prudent general opposes.[92] It is also noticeable that Kallimachos' second brother argues that the high mountain with rocky terrain and tall trees is unsuitable for the crossing of a large army followed by a large baggage train made up of camels. He also suggests that the army should not march through an uninhabited area.[93] The heroic general Belisarios, equally, executed one of his subordinates who complained that the burning of the fleet from which they landed in *Inglittera* was not a proper military action.[94] That the romances often specify that the heroic acts they describe are not compatible with prudent military thinking shows that the authors of these works possessed sound military knowledge and knew that the deeds they mention had little to do with battlefield realities.

---

[89] The battle of Pelagonia is very well documented and discussed in detail by modern scholarship. See Akropolites, I, 168–171; Pachymeres, I, 121–123; Gregoras, I, 71–75; *Chronicle of the Morea*, vv. 3550–3789; D. Nicol, *The despotate of Epiros* (Oxford, 1957), 170–182; D. Geanakoplos, *Emperor Michael Palaiologos and the West. A study in Byzantine-Latin Relations. 1258–1282* (Harvard, 1959), 47–74; J.V.A. Fine, *The Late Byzantine Balkans* (Kalamazoo, 1995), 162–165; Shawcross, *Chronicle of the Morea*, 73–76; R. Milhajlovski, "The Battle of Pelagonia, 1259: a New Look at the March Routes and Topography," *BSI* 64 (2006), 275–284.

[90] Kantakouzenos, I, 90–92.

[91] *Achilleid*, vv. 460–464.

[92] *Kallimachos*, vv. 229–230.

[93] *Kallimachos*, vv. 89–101.

[94] *Ιστορία του Βελισαρίου* [*The Story of Belisarios*], W. Bakker – A. van Gemert (eds.), (Athens, 1988), v. 202.

At the same time, the sources provide a considerable number of examples of generals blamed for imprudent action and miscalculation of the conditions in which they engaged in battle. In 1255, Constantine Strategopoulos and Constantine Tornikes, heading from Serres to Tzepaina, 'proved to be bad generals,' according to Akropolites. They fled before estimating the strength of the attacking Bulgarians and they left a large amount of booty in the hands of the enemy. Akropolites provides a justification, claiming that they were leading poor troops.[95] He also provides a number of other examples of bad generalship which led to defeats. He claims that Manuel Laskaris imprudently attacked a group of Mongol mounted archers leading heavy cavalry troops.[96] Similarly, Akropolites blames Xyleas for an attack on raiding Serbian troops without having scouted the area and gathered the necessary intelligence.[97] Regardless of whether Akropolites is to be believed or not, it should not be forgotten that he takes a negative view of those promoted by Theodore II to high military posts, these examples are indicative of what was considered bad leadership by the Byzantines. Furthermore, in 1347 the troops of the Phakrases defeated those of Dobrotič, because, as Kantakouzenos states, Dobrotič proved to be a bad general by sending his cavalry to fight in vineyards and ditches.[98]

The emphasis that the later Byzantines put on avoiding battle does not differ from Western European attitudes at the time. The *Chronicle of the Morea*, despite its inaccuracies, reflects vividly the differences in battlefield tactics adopted by the Byzantines and the Latins in the Peloponnese. This source is negative towards the Byzantines and insists that they avoided fighting the Latins in the open field stating that the Byzantines 'would not fight the Latins with lances but with bows,' to demonstrate that the Byzantines were cowards in contrast to the brave Frankish warriors, who sought to fight in close quarters. However, this conduct, which was labeled deceitful when applied to the Byzantines, received a different treatment when the perpetrators were the Latins. Being aware of the fact that such views about bravery have little to do with the realities of warfare, the author of the *Chronicle* admits

[95] Akropolites, I, 114; Theodore II, *Epistulae*, 252.

[96] Akropolites, I, 123. For Manuel Laskaris see *PLP*, 14551.

[97] Akropolites, I, 145.

[98] Kantakouzenos, II, 585; George Phakrases was *protostrator* from 1346 to 1355. He was one of the generals of Kantakouzenos' army in Didymoteichon. See *PLP*, 29575. Dobrotič was despot of Dobrutza: *PLP*, 29073.

that, 'a brave soldier should combine resourcefulness and prudence and should fight cautiously against his enemies because as the prudent rightfully say, craft and cunning win over bravery.'[99]

This similarity in attitudes to warfare is reflected in the treatise compiled by Theodore Palaiologos, the Byzantine ruler of an Italian principality. He comments that if the enemy is stronger, then it is impossible to fight a pitched battle. He also suggests that if an enemy is attacking and ravaging the lands of the ruler, the ruler should try to use spies to collect information about the strength and movements of the enemy.[100] Moreover, the marquis of Montferrat refers to the usefulness, as well as the impact, of raiding warfare and to the need to avoid fighting against a seemingly stronger enemy.[101] It seems that even in Western Europe the image of knights as illiterate and undisciplined soldiers who love war is exaggerated and modern scholarship has modified the popular view promoted by chivalric literature of the knight dominating warfare as an individualistic soldier. The battlefield reality was quite different, and the tactics employed by Western Europeans required high levels of personal discipline, as well as the ability of the knights to operate effectively in groups.[102] Similarly, although heroism was praised, the Byzantine attitudes to warfare did not favour individualism. Moreover, discussing the military treatises compiled in fifteenth-century Italy, Michael Mallett concludes that these works emphasise the need for prudence and caution and epitomise the basic concept of fourteenth and fifteenth-century Italian and Western European warfare, which was as far as possible defensive and counteroffensive and relied on stratagem and deceit. Western European commanders were cautious about engaging the armies in major confrontations and were aware of the cost and risk of major military enterprises. [103]

[99] *Chronicle of the Morea*, vv. 4930–4932; Shawcross, *The Chronicle*, 196–197, 201.

[100] Palaiologos Theodore, marquis of Montferrat, *Les enseignements de Théodore Paléologue*, C. Knowles (ed.), (London, 1983), 89–90.

[101] Theodore Palaiologos, *Les enseignements*, 80–82.

[102] B.S. Bachrach, "Caballus and Caballarius in Medieval Warfare," in H. Chikering – T.H. Seiler (eds.), *The Study of Chivalry. Resources and Approaches* (Kalamazoo, 1988), 197.

[103] M. Mallett, *Mercenaries and their Masters. Warfare in Renaissance Italy* (London, 1974), 177; France, *Western Warfare*, 13–16; Keen, *Chivalry*, 221–222.

The ideal leadership

Despite the prevalence of a cautious attitude to battle, the sources show that a number of generals had developed a reputation for their courage, bravery and generosity towards their troops. Pachymeres states that John Palaiologos achieved great deeds, and was feared by everyone for both his reputation and stature. Gregoras calls him unrestrained in his assaults, but prudent.[104] He also reports that Michael Doukas Glavas Tarchaneiotes was so experienced in the art of war that the other generals were like children before him.[105] Similarly, Gregoras states that Michael Palaiologos had the prestige required by a ruler, due to long experience and practice in war.[106] He also relates that Manuel Tarchaneiotes was a close relative of the emperor and carried out many acts of heroism. He sustained many wounds in wars, he fought many battles and for this reason he acquired great experience in the art of war.[107] Kantakouzenos praises Manuel Tagaris for showing great courage and achieving great things against the Turks in Philadelphia.[108] Perhaps the most characteristic example of the influence of reputation is the *pinkernes* Alexios Philanthropenos. According to Gregoras, Philanthropenos' successful campaign against the Turks in 1295 had such an impact that, when he was sent to Philadelphia in 1323 without an army, either due to the lack of resources or because the emperor did not trust him, the Turks retreated.[109] It is quite possible that his reputation was the main reason for his appointment by Andronikos III as governor of the Aegean islands under Byzantine control in the 1330s.[110] The importance of charismatic military leaders is accentuated by the fact that Alexios Philanthropenos was blind. Pseudo-Lachanas and Choirovoskos provide two further examples of charismatic military leadership. In 1294 and in 1304, respectively, in a period when the Byzantine army could not resist the Turcoman chiefdoms, they managed to raise troops in Asia Minor to fight the Turks.[111]

---

[104] Pachymeres, I, 151, 285; Gregoras, I, 107.
[105] Gregoras, I, 159.
[106] Gregoras, I, 153.
[107] Gregoras, II, 652.
[108] Kantakouzenos, I, 90.
[109] Gregoras, I, 369.
[110] Gregoras, I, 534; Kantakouzenos, I, 488–494.
[111] Pachymeres, III, 211–213, IV, 487–489. In March 1294, Pseudo-Lachanas succeeded in organising a significant force in Mesothynia. Pachymeres states that it was

The Byzantine commanders secured the loyalty and trust of their troops by their generosity, and by being on friendly terms with them, rather than by impressing them with their bravery and military deeds. The available source material indicates that the most effective way to maintain high levels of discipline and morale was to be close to the common soldiers, to encourage them by speeches, financial rewards, and, more importantly, hopes of the prospect of material profit. The sources unanimously connect financial reward and a friendly relationship between troops and leadership with the boosting of morale and military success. Pachymeres states that, even before Michael VIII ascended the throne he had taken special care of the frontier troops in Philadelphia, and by offering gifts, he inspired them to be more eager guards and more confident in their hopes.[112] Similar virtues are attributed to John Palaiologos, who, according to Pachymeres, surpassed everyone in generosity and himself detested money. As head of the army, John Palaiologos took care of the troops by speeches and gifts and above all treated his soldiers not as a despot, but as a brother.[113] Alexios Philanthropenos' success in Asia Minor is attributed to his generosity not only towards his troops but also towards the enemy, thus making their surrender to the Byzantines a more attractive option and, as Pachymeres states, reducing their 'raging madness against us.'[114] Praising the qualities of Alexios Philanthropenos, Gregoras states that he was generous and affable to his subordinates. He points out that this is the best means for a general to achieve victories and trophies.[115] Planoudes, who claims that he witnessed the vast quantity of spoils captured by Philantropenos' army, confirms Pachymeres and Gregoras by mentioning the Achyraitai, a group which hoped to be benefited by Philanthropenos' successes.[116] Kantakouzenos provides a more specific

composed of farmers, ignorant of the needs and realities of warfare, and their conduct was that of bandits, not of soldiers. In the summer of 1304, John Choirovoskos, leading a band of 300 archers and mace-bearers, offered to fight in Asia Minor against the Turks. Initially, he was arrested by the imperial authorities, but he escaped in Asia Minor, where he organised a force on the Skamandros. Eventually he was defeated at the hands of the Turks. He reappears in 1307, when he was sent to lead a force of 1000 in Macedonia against the Catalans and their Turkish allies. However, this force did not achieve anything.

112 Pachymeres, I, 141.

113 Pachymeres, I, 285.

114 Pachymeres, III, 239; Gregoras, I, 196.

115 Gregoras, I, 196.

116 Pachymeres, III, 239; *Maximi Monachi Planudis Epistulae*, A. Leone (ed.), (Amsterdam, 1991), 116–118, 150, 154, 171–172, 180–181.

example. In 1350, in the battle on the walls of Edessa, Kantakouzenos promised a reward of four *mnas* to whoever would be the first to bring down a banner from the walls, three *mnas* to the second and two *mnas* to the third.[117] Similarly, Gregoras states that Syrgiannes secured the loyalty of his troops by his generosity, in particular by providing his troops with much booty.[118] The same historian states that on the eve of the civil war of 1341–1347, the army supported Kantakouzenos because,

> They saw him as their wealth provider and as a fellow soldier rather than their general. He used to share the pains of the campaigns with them encamping in the countryside even during the winter.[119]

Suggestions of this kind can also be found in the military manuals of the earlier periods. The *Taktika* of Leo suggests that generals should reward soldiers and units which excelled themselves on the battlefield with armour, booty and offices.[120] Moreover, generosity towards brave soldiers is an ideal promoted by works of fiction. The anonymous author of the *Story of Belisarios* describes how his protagonist, the heroic general Belisarios, rewarded the soldiers who were the first to raise the imperial standards on the enemy walls with two splendid horses, beautiful saddles and golden swords.[121]

The military manuals of the middle Byzantine period indicate that military leaders inspired their troops by battle orations. The *Taktika* of Leo the Wise suggests that the general should be a good orator, who raises the spirits of the soldiers for the battle and boosts their morale.[122] The author of the treatise *On Skirmishing* admonishes the commander to inspire the soldiers by delivering speeches beginning with the phrase 'we are standing in a manly and brave way.'[123] By talking to his troops, the military leader wanted to show that he was not different from his soldiers, that he cared for their welfare, and that they should see him as comrade or father, rather than as the unapproachable leader of a higher social status who did not share their interests. These

---

[117] Kantakouzenos, III, 129. It is not clear what currency Kantakouzenos had in mind.

[118] Gregoras, I, 297.

[119] Gregoras, I, 586.

[120] *Taktika*, XVI. 4.

[121] *Ιστορία του Βελισσαρίου*, vv. 172–173.

[122] *Taktika*, II. 12.

[123] *Le traité sur la Guérilla (De velitatione) de l'empereur Nicéphore Phocas (963–969)*, G. Dagron - H. Mihăescu (eds.), (Paris, 1986), 125.

principles continued to apply in the late period. In his encomium of Andronikos II, Choumnos comments that the emperor enhanced the fighting spirit of the army of Asia through speeches.[124] Encouragement through orations was more important when the army was in a very difficult position and defeat seemed imminent. According to Gregoras, before the battle of Rossokastro (1332) Andronikos III delivered a speech to encourage his troops, exhorting them not to be afraid since small armies very often defeated larger forces.[125] Similarly, in 1343, when the army of Kantakouzenos was encircled by Apokaukos and the Serbians in the Axios river, he delivered a speech reminding his soldiers that in the past Byzantine emperors had managed through bravery to achieve a victory against the odds.[126] Another interesting example is provided by the *Story of Belisarios*. The hero of this work, the general Belisarios, tries to encourage his soldiers by calling them 'brothers.'[127] The accuracy of the general's words may be questioned, since these speeches are rhetorical products of the authors of these historical accounts. Nonetheless, it has been argued that the tradition of historiography demanded that devices of amplification and ornamentation such as speeches had to be plausible and consequently, that these orations reflect the author's conceptions of motivation and morale in war.[128]

The picture of ideal generalship, which includes, as discussed above, all the skills a general needed, is provided by Gregoras. Gregoras portrays Philes Palaiologos as the ideal Byzantine military commander who in 1313 managed to restore the morale of the forces under his command and to defeat the Turks of Halil, who were plundering Thrace after breaking from the Catalan Grand Company which had followed as allies in their advance towards Thessaly and Attica. Philes Palaiologos was not a full-time 'professional' military commander. Gregoras implies that he owed his success to his attitude to his troops. He had realised that the inability of the state to resist the Turks of Halil in Thrace had damaged the morale of the Byzantine forces. He requested permission from Andronikos II to lead the Byzantine

[124] *AG*, I, 21.
[125] Gregoras, I, 486.
[126] Kantakouzenos, II, 251.
[127] *Ιστορία του Βελισαρίου*, v. 190.
[128] See R.E. Bliese "Rhetoric and Morale: A Study of Battle Orations from the Central Middle Ages," *JMH* 15 (1989), 201–226.

troops against Halil, asking not only for the command of the troops but also the freedom personally to select his soldiers. This means that he probably chose those he considered more loyal to him. According to Gregoras, Philes did not have the figure of a military man; he was thin and weak and most of the time he was ill. Despite his bodily weakness, however,

> He ignited the flame of war in his troops by his friendliness towards them and by donating horses and money. He even gave his belt to one soldier and to another he gave the knife he wore beside his sword. He lived and slept with his soldiers, promising them high honours and grants after the war according to the performance of each man. Before the battle, he tried to encourage them through speeches.[129]

Although there are doubts concerning the reliability of Gregoras' account, for he does not mention the impact of the reinforcement by a force of 2000 soldiers from the Serbian ruler Stefan Uroš II Milutin (1282–1321), this description of Philes' role is indicative of what the late Byzantines perceived as an ideal general.[130]

---

[129] Gregoras, I, 263–266.
[130] Gregoras, I, 268.

CHAPTER THREE

SOLDIERS

Pronoia *holders*

The late Byzantine state employed a variety of methods to finance its military forces. Late Byzantine soldiers were maintained through grants of land, cash, tax exemptions, rights to state revenues. That often soldiers were remunerated through a combination of the above-mentioned methods makes impossible any rigid categorisation of the military forces employed by the Byzantine rulers. For instance, there were mercenaries, who received *pronoia* grants. The categorisation of soldiers becomes more difficult due to the lack of sufficient sources. Almost nothing is known about the financing of the Cumans enlisted by John III in 1240 and quite often, it is not clear whether the land possessed by soldiers was related to their military service. Taking into account these developments the Byzantine soldiers can be divided into two broad categories: To landholders, who received their remuneration in the form of grants of land, and to mercenaries, who were paid in cash. However there were significant overlaps.

The sources show that a large part of the army was maintained through *pronoia* grants. It would be superfluous to discuss in detail the function and significance of the institution of *pronoia*.[1] However, it is useful to highlight some important points concerning the importance and impact of *pronoia* on the late Byzantine military organisation. *Pronoia* was a grant by the emperor of the state's fiscal rights over revenue sources to an individual or group of individuals.[2] The extensive

[1] For a detailed discussion of the institution of *pronoia* and a comparison with the fief, *Iqta* and *timar* see Bartusis, *The Late Byzantine Army*, 162–190; See also N. Oikonomides, "Contribution à l' étude de la *pronoia* au XIIIe siècle," *REB* 22 (1964), 158–175; idem, "The Role of the Byzantine State in the Economy," in A. Laiou (ed.), *The Economic History of Byzantium* (Washington, 2002), 1042–1050; idem, "The properties of the Deblitzenoi in the Fourteenth and Fifteenth Centuries," in A. Laiou (ed.), *Charanis Studies: Essays in Honor of Peter Charanis* (New Brunswick, 1980), 176–198; A. Kazhdan, "*Pronoia*: The History of a Scholarly Discussion," *MHR* 10 (1995), 133–163.

[2] J. Haldon, *The State and the Tributary Mode of Production* (London, 1993), 130; Oikonomides, "The Role of the Byzantine State," 1048; A. Laiou, *Peasant Society in the*

use of the institution of *pronoia* reflects the inability of the Byzantine state to exact resources effectively. As a result, the state had to rely on the consumption of its resources by its troops and administrators directly.[3] Some *pronoiai* consisted the entire payment to the soldiers and others only part of it.[4] Moreover, the size of *pronoia* grants varied significantly. There were large *pronoiai* amounting to hundreds of *nomismata*, e.g. Pachymeres reports that the value of Constantine Palaiologos' (Michael VIII's son) *pronoiai* was 60,000 *nomismata*, whereas other *pronoiai* were so small that the social status of their holders was not much higher than that of a group of peasants.[5] In conclusion, *pronoia* was the means of financing of both wealthy and smallholding soldiers.

From the administrative point of view, the *pronoia* soldier was inexpensive for the Byzantine state. He was expected to be permanently in the service of the state, which would provide him with the 'immobile' means of his subsistence.[6] From the military point of view, the *pronoia* system was of important significance. The prospect of a reward in the form of *pronoia* would have significant impact on the motivation and effectiveness of the individual soldier, who would not only receive significant income but could also afford better equipment as an eager defender of the source of his income. The *prostagma* Michael VIII issued in 1272 to define Andronikos II's position and rights as co-emperor, is indicative of this.[7] It stipulates that if a soldier appears useful, he may receive an *oikonomia* (*pronoia*) increased by 24–26 *hyperpyra*. In the same manner, if the soldier is not conducting himself well, he loses his *oikonomia*, which should be granted to a more deserving soldier. It seems that, following the principles of this *prostagma*, in 1293 Andronikos II accepted the request of Leo Koteanitzes to receive land in the area of Preasnitsa in Macedonia,

---

Late Byzantine Empire: A Social and Demographic Study (Princeton, 1977), 6; On the collective *pronoiai* see Oikonomides, "À propos des armées," 367–368.

[3] Haldon, *Tributary Mode,* 198.

[4] Oikonomides, "The Role of the State," 1047.

[5] Pachymeres, III, 175; Bartusis, *The Late Byzantine Army,* 174–175; Oikonomides, "À propos des armées," 354, 362–364; *Actes de Docheiariou*, N. Oikonomides (ed.), (Paris, 1984), 138–143, 161–162, 234.

[6] Oikonomides, "À propos des armées," 355.

[7] A. Heisenberg, *Quellen und Studien zur spätbyzantinischen Geschichte* (London, 1973), 40–41.

as a reward for his loyalty and for his exceptional performance in the defence of the empire against 'enemy attacks.'[8]

Theodore I and John III are said to have used the *pronoia* system extensively, increasing the size and effectiveness of the army.[9] According to Pachymeres, when Michael VIII ascended the throne, he provided the soldiers with daily grants and transformed the *pronoiai* to hereditary property, aiming at making the soldiers more eager fighters and at securing their loyalty. Similarly, he awarded grants to the frontier troops in Philadelphia and showed special care for its fortifications, trying to make the border guards more willing and brave soldiers.[10] Writing between 1296 and 1303 the orator Nikolaos Lampenos praises Andronikos II for making the *pronoiai* hereditary not only to sons of soldiers but to every inheritor.[11] Unless Lampenos does not say the truth, it is possible that this measure was a temporary one because in about 1304, Thomas Magistros urged the emperor to rely on native soldiers and to provide them with *pronoiai*, which he should make hereditary for at least one generation for families of soldiers killed in battle. Moreover, Magistros comments that soldiers should have properties to defend. This would make them better warriors who would be ready to fight and die for their properties and the emperor whom they would regard as their benefactor.[12]

In a difficult to interpret text, Pachymeres, referring to the Turkish advance in Asia Minor in 1303, states that the government considered the confiscation and distribution to the local population of *pronoiai* held by ecclesiastical institutions, as well as by wealthy individuals in Asia Minor, who had abandoned their land. The government by doing so expected that the new *pronoia* holders would be motivated to fight for their properties.[13] In a similar example, Pachymeres states that when in 1304 the Bulgarians raided Thrace, Michael IX was expected to arrive there soon and defend his new *pronoiai*, which he received

---

[8] *Actes de Chilandar*, M. Živojinović – V. Kravari – C. Giros (eds.), (Paris, 1998), I, 147.

[9] *Synopsis Chronike*, 507–508; H. Ahrweiler, "La politique agraire des empereurs de Nicée," *B* 29 (1958), 57.

[10] Pachymeres, I, 139–141.

[11] Lampenos, *Encomion*, 68; Angelov, *Imperial Ideology*, 137.

[12] *PG* 145, cols. 461–464.

[13] Pachymeres, IV, 425–427; For the dating of these events see A. Failler, "Pachymeriana alia," *REB* 51 (1993), 248–258.

in compensation for those he had lost in Asia Minor.[14] Therefore, connecting the soldiers to the land would make them more eager defenders, since they would have the incentive to fight for the source of their personal income. Consequently, *pronoia* grants played an important role in the motivation of soldiers.

The *exisoseis*, the assessments of *pronoiai*, were the means by which the state granted *pronoiai*. The aims of the assessment were to provide the soldiers with the means to afford to arm and maintain themselves and by doing so, to increase the size of the army improving its effectiveness and morale. However, the need for *exisoseis* reflects the existence of administrative inefficiencies and the openness of the *pronoia* system to abuse, as well as the lack of financial resources, which inevitably resulted in the reduction of the effectiveness of the army. In 1298, John Tarchaneiotes was sent to Asia Minor to reform the military administration of the area. He found that many soldiers had lost their *pronoiai*, while others who did not perform military service had increased theirs by bribing the local authorities. The soldiers who lost their *pronoiai* could not perform military service, since they were deprived of the necessary resources to maintain and arm themselves.[15] In 1327, after the victory of the *protostrator* Theodore Synadenos over Constantine Asan at the Melanas river, which was crucial for the outcome of the civil war between the two Andronikoi, Andronikos III, in Kantakouzenos' words,

> Established assessors in Thessalonica, who carried out a fiscal reassessment of the western parts of the state and established new (military) rolls in addition to the ones in existence aiming at increasing the size of the army.[16]

This statement reflects the reduction of the army, its disorganisation during the civil war and the attempt of Andronikos III to secure its loyalty. In another example indicating the continuous need for reforms, as well as the confidence of the Byzantine government in the effectiveness of the connection between land and maintenance of the army, Manuel II Palaiologos, as despot of Thessalonica, responded to the emergency situation caused by the Serbian defeat at the hands of

---

[14] Pachymeres, IV, 447.

[15] Pachymeres, IV, 285; Bartusis, *The Late Byzantine Army*, 75; Oikonomides, "The Role of the State," 1044.

[16] Kantakouzenos, I, 287–288; Gregoras, II, 595.

the Ottomans at the Maritza river in 1371, by confiscating ecclesiastical lands to fund his army.[17]

Kantakouzenos' account on the reforms he carried out in 1341 in order to secure the support of the army in his attempt to seize the throne, reveals the aims and ideal effects of the financial assessment. He writes that,

> Not only many of those registered in the military rolls (*katalogoi*), but also not a few of the rest neglected their military duties, on the grounds that the salaries received from the emperor had been discontinued.

Therefore, he was contemplating ways to relieve the soldiers and to restore the *pronoiai* they received from the emperor. The task was assigned to an experienced assessor, Patrikiotes, who in the past had taken advantage of his position, enriching himself through the fiscal reassessments he carried out. Kantakouzenos, stressing the role of the fiscal assessments states that as the judge's task is to take care of the good order in the cities, and to impose the law, the assessor, if he wants to deliver justice and to serve the common good, takes away from those who possess something more than is fair and adds to those who have less. By doing so, the assessor makes everyone willing to campaign, to be disciplined and obedient to the generals.

Kantakouzenos affirms that when the process of the reassessment was completed, the soldiers expressed their gratitude to him and became willing to fight against the enemies of the Byzantines in every part of the Byzantine land. Furthermore, due to their increased incomes the soldiers were able to use better weapons and they could afford better and more war horses than before.[18] In an earlier instance, Pachymeres indicates the importance of *pronoia* for the maintenance of the army. He relates that many Byzantine soldiers had lost their *pronoiai* in Asia Minor as a result of the expansion of the Turcoman principalities and had fled to the west at the beginning of the fourteenth century. Many of them were experienced and good soldiers who were unarmed, because they had been deprived of their possessions.[19] The army of Michael IX which in 1305 was defeated in the battle of Apros at the

[17] Barker, *Manuel II*, 17; K. Smyrlis, "The State, the Land, and Private Property: Confiscating Monastic and Church Properties in the Palaiologan Period," in D. Angelov (ed.), *Church and Society in Late Byzantium* (Kalamazoo, 2009), 66–72.

[18] Kantakouzenos, II, 58–62.

[19] Pachymeres, IV, 459.

hands of the Catalans, was partly composed of such Anatolian soldiers, who seem to have been completely demoralised, ineffective and lacking sufficient weapons, although just a few decades earlier they were the heart of the army that succeeded in recovering significant parts of the old Byzantine empire.[20]

It is important to stress the similarities between Kantakouzenos' and Pachymeres' accounts concerning the connection of the morale, size and effectiveness of the army to the reforms of the *pronoia* system. However, none of the above-mentioned reforms had long-term results. Tarchaneiotes faced strong reaction from the Anatolian elite, who had enriched themselves, and from the church, which did not accept a supporter of the Arsenite movement in a high post.[21] Andronikos III initially had to deal with dissension and conflicts within the army, which were the result of the eight-year civil war and possibly affected the outcome of the battle of Pelekanos-Philokrene in 1329, where the Byzantines were defeated at the hands of the Ottomans.[22] When Andronikos III managed to organise his army, it proved to be insufficient, in terms of manpower and resources, for the length of the frontiers and the number of enemies. The improvement of the effectiveness of Andronikos III's army could not match the pace of the increasing effectiveness, both in terms of combat strength and administration, of the forces of the Turcoman emirates, mainly of the Ottomans. As far as Kantakouzenos' reforms are concerned, any positive result was cancelled by the outbreak of the civil war in 1341, although most likely many of the soldiers who benefited from these reforms fought alongside Kantakouzenos.

These failed reforms illustrate the ineffectiveness of the late Byzantine military administration. Moreover, the inability of the throne to defend their properties against enemy raids and the corruption of local officials caused widespread dissatisfaction among the

[20] Pachymeres, IV, 599.

[21] In 1265, a synod deposed the patriarch Arsenios who had excommunicated Michael VIII Palaiologos on account of his treatment of the young John IV Laskaris. Many clergymen and laymen refused to recognise Arsenios' successors and formed a schism that would last for almost half a century. The Arsenite schism is connected to the dynastic opposition to the Palaiologoi, as well as to the hostility of the population of Asia Minor to the Palaiologoi: Pachymeres, III, 285–289; Geannakoplos, *Emperor Michael*, 271–272; Laiou, *Constantinople and the Latins*, 20–23, 34–36; P. Gounarides, *Το κίνημα των Αρσενιατών (1261–1310)* [*The Movement of the Arsenites*] (Athens, 1999), 69.

[22] Kantakouzenos, III, 363.

soldiers leading many to abandon the Byzantine army. According to Pachymeres, corruption and the pressure of the Turkish raids induced many Byzantine soldiers in Asia Minor to join the enemy. He remarks that shortly after the recovery of Constantinople in 1261, the rapacity of local generals who kept the lion's share of the booty captured in wars for themselves induced Byzantine soldiers to join the Turks. Discussing the political and military situation in Paphlagonia in the second half of the thirteenth century, Pachymeres remarked that the locals became allies and guides of the Turks, having considered the benefits of this act.[23] By joining the Turks, local soldiers protected their estates from enemy raids. Furthermore, they must have been attracted to the wealth they could gain through plunder and the low taxation imposed by the Turcoman principalities. Enrichment through plunder and conquest could bring together Byzantine and Turkish warriors.[24] In addition, the available evidence suggests that Byzantine pronoiars were incorporated to the Ottoman *timar* system, which was similar to the institution of *pronoia*.[25] Military groups from the annexed states were incorporated into the Ottoman army. The fact that military men from the conquered states continued to perform their duties under the new regime satisfied the need of the Ottomans for manpower. It has been concluded that at least one-third of the fifteenth-century Balkan *timariots* were Christians.[26]

Furthermore, Gregoras attributes the initial success of Syrgiannes' revolt in Macedonia in the 1330s to the fear of the local population that the rebels would destroy their harvest and to promises of *pronoia* grants, cash and offices.[27] Promises of grants of privileges facilitated the Serbian expansion in Macedonia, Thessaly and Epiros in the 1340s, when Byzantium was plunged into a catastrophic civil war.[28] The

---

[23] Pachymeres, I 35, 293.

[24] See H. Lowry, *The Nature of the Early Ottoman State* (New York, 2003), 46–57.

[25] H. İnalcık, "Ottoman Methods of Conquest," *Studia Islamica* 2 (1954), 113–114; Lowry, *Early Ottoman State*, 90–92; For an analysis of the *timar* system see N. Beldiceanu, *Le timar dans l' État ottoman (début XIVe–début XVIe siècle)* (Wiesbaden, 1980), passim.

[26] H. İnalcık, "Stefan Duşan' dan Osmanlı imperatorluğuna," [From Stefan Dušan to the Ottoman Empire] in idem, *Fatih devri üzerinde tetkiler ve vesikalar* [Researches and Documents on the Era of the Conqueror] (Ankara, 1954), 141–146, 170, 182–183; Lowry, *Ottoman Balkans*, 7–8.

[27] Gregoras, I, 495, 499.

[28] G. Soulis, *The Serbs and Byzantium during the Reign of Tsar Stephen Dušan (1331–1355) and his Successors* (Washington, 1984), 87.

instability caused by the civil war of 1341–1347 and the hope of receiving privileges from the Serbians, who at the time were the strongest power in the Balkans, explain why the Byzantine troops in Macedonia did not put up any serious resistance to the Serbian army. Moreover, by adopting the title of emperor of Serbians and Romans Stefan Dušan facilitated the entrance of Byzantine troops to his service and undermined the authority of the emperor in Constantinople.[29] It was reasonable for the Byzantine soldiers in Macedonia to join the Serbians in order to secure their holdings and increase their property by taking advantage of the expansion of the Serbian state. For instance, throughout the first half of the fourteenth century the Devlitzenoi are soldiers fighting for Byzantium and in 1383, one of them died in Chortiates where the Ottomans defeated the army of Manuel II. Shortly afterwards a Devlitzenos, who apparently was a member of the same family, is called the 'Serbian.'[30]

The Allagia

The organisation of soldiers maintained through *pronoia* is a complicated issue. The main problem is the lack of sufficient information and the problematic nature of the available sources. Late Byzantine authors use inconsistent and archaic military terminology. Nevertheless, the available sources connect the *pronoia* soldiers to military units, which were stationed in provinces and cities, without, however, being part of the local garrisons. The first example suggesting that the troops stationed in the provinces were put under a different command from the local garrisons comes from the Nicaean period. When in 1255, Dragotas, the head of the army of Melnik, rebelled he assembled around him all the soldiers of Melnik, as well as many others, as Akropolites states, and besieged the city. The heads of the garrison of Melnik were Theodore Nestongos and John Angelos.[31] Although there is no evi-

[29] L. Maksimovič, "Η ανάπτυξη κεντρόφυγων ροπών στις σχέσεις Βυζαντίου-Σερβίας τον 14ο αιώνα," [The Development of Centrifugal Forces in the Byzantine-Serbian Relations in the Fourteenth Century] in N. Oikonomides (ed.), *Byzantium at War* (Athens, 1997), 286.

[30] D. Jacoby, "Foreigners and the Urban Economy in Thessalonike c. 1150–c. 1450," *DOP* 57 (2003), 87; P. Schreiner, "Zwei unedierte Praktika aus der zweiten Hälfte des 14.Jh.," *JÖB* 19 (1970), 34.

[31] Akropolites, I, 114–115.

dence to suggest that Dragotas' army was made up of pronoiars, the sources of the Palaiologan period provide convincing evidence that the armies stationed in the provinces were made up of *pronoia* holders. The *chrysobull* Andronikos II issued in 1319 by which he granted the city of Ioannina various privileges, stipulates that,

> The *kastrenoi* (urban population of Ioannina) are not obliged to serve in military campaigns away from Ioannina; only the soldiers who belong to the ranks of *allagia* and hold an *oikonomia* have to serve.[32]

This document indicates that, the units made up of *pronoia* holders who were stationed in the provinces, were called *allagia*. They reinforced the local defence and, unlike the *kastrenoi*, they participated in military campaigns.[33] Furthermore, Michael VIII's *prostagma* issued in 1272 for his son connects the *pronoia* holders with the *allagia*.[34] However, any discussion concerning the *allagia*, should take into account the inconsistent terminology employed by the sources. The term *allagia* is not the only one to refer to military units composed of pronoiars. Apart from the *allagia* the sources mention the *vasilika* (imperial) *allagia*, the *vasilike* (imperial) *taxis* the *megala* (great) *allagia*, the *tagmata*, the *authentikon allagion* implying the existence of more than one type of *allagion*. It is almost impossible to reconstruct the internal organisation of these units. The limited available sources show that they would fight on the battlefield under the orders of the *kephale*, the provincial governor who was in charge of both the civil and military administration of the locality he was appointed to administer. In this context, the role of the late Byzantine *kephale* was different from that of the middle Byzantine *strategoi* (generals) of the *themes*, since the pronoiars were relatively independent from the local administration.[35]

An important question is what the differences were between all the above-mentioned types of *allagion*, which were made up, at least

---

[32] MM, V, 81; For an analysis of this document see Kyritses, "Common Chrysobulls," 237–238; Nicol, *The Despotate of Epiros. 1267–1479*, 83–89.

[33] B. Hendrickx, "Allagion, Tzaousios et Protallagatôr dans le context Moréote: quelques remarques," *REB* 50 (1992), 208.

[34] Heisenberg, *Quellen*, 40.

[35] L. Maksimović, *The Byzantine Provincial Administration under the Palaiologoi* (Amsterdam, 1988), 147. For the possibility that the *megas tzaousios* was in charge of the *megala allagia* see M. Bartusis, "The Megala Allagia and the Tzaousios: Aspects of Provincial Military Organisation in late Byzantium," *REB* 47 (1989), 183–205; Hendrickx, "Allagion," 208–217.

largely, of *pronoia* soldiers. The term *allagion* was not new in late Byzantium. According to Haldon,

> During the later ninth and tenth centuries and afterwards the term *allagion* was applied to a unit of soldiers of varying size, and to those regiments of the watch whose duties involved changing place at regular intervals and relieving one another.[36]

This term appears also in the military treatises of the middle Byzantine period. In the tenth-century manuals entitled *Sylloge Tacticorum*, and *Praecepta militaria* the *allagion* is defined as a cavalry unit. Moreover, the *Sylloge Tacticorum* mentions the existence of *vasilika allagia* stationed in the provinces of Thrakesion and Charsianon in Asia Minor.[37]

In late Byzantium, the term *allagion* appears repeatedly in the *Chronicle of the Morea*, in the *War of Troy* and in the *Achilleid* to refer to cavalry, and possibly to infantry units, which on the battlefield were divided into *syntaxeis*.[38] Describing the army John Palaiologos led on his campaigns in Europe in the 1260s Pachymeres writes that, 'the army was composed of many *allagia*, as those who talk about *tagmata* would say.' He mentions the *allagia* of Paphlagonia, Mesothynia, Thrace, Neokastra, Macedonia, Mylassa, as well as the Cumans and the Latins.[39] Therefore, Pachymeres specifies that *allagia* and *tagmata* are the same thing. This connection between *tagmata* and *allagia* is confirmed by Pseudo-Kodinos, who states that the army was divided into *tagmata* stationed in the provinces and that the *megaloallagitai* are parts of the *tagmata*.[40] The identification of the *allagia* on the basis of the province in which they were stationed means that the *allagia* were not only squadrons deployed on the battlefield, but also units stationed in the provinces.

As has been stated, there were various types of military units called *allagia*. The term *vasilikon allagion* is used by the sources to refer to

---

[36] J. Haldon, *Byzantine Praetorians* (Bonn, 1984), 275.

[37] Haldon, *Praetorians*, 277; *Sylloge tacticorum, quae olim 'inedita Leonis Tactica' dicebatur*, A. Dain (ed.), (Paris, 1938), 56.

[38] Aerts and Hokwerda argue that the term *allagion* in the *Chronicle of the Morea* applies not only to cavalry but also to infantry units: W.J. Aerts - H. Hokwerda, *Lexikon of the Chronicle of the Morea* (Groningen, 2002), 14; *Achilleid*, vv. 399–402, 440, 469–473, 483, 509–524, 545–550, 568–569; *The War of Troy*, vv. 918, 922,928,4381; Hendrickx, "Allagion," 208–210.

[39] Pachymeres, II, 403.

[40] Pseudo-Kodinos, 187.

two different things. To the imperial escort and to mobile units of the army stationed in the provinces. The presence of the imperial escort, which was called *vasilikon allagion*, on campaigns led by the emperor is attested in three cases. In 1255, during his campaign against the Bulgarians Theodore II sent his uncle Manuel Laskaris and the head of the *vasilikon allagion*, Constantine Margarites, to scout the area around Vatkounion. According to Akropolites, Margarites had previously served in the army of the theme of Neokastra; later he was promoted to *tzaousios* and Theodore II made him *megas tzaousios* and head of his *taxis*.[41] In 1329, in the battle of Pelekanos-Philokrene Andronikos III was followed by his own *tagma*, led by Manuel Asan.[42] The same unit under the same leader appeared also in 1332 in the battle of Rosokastron between the Byzantines under the command of Andronikos III and the Bulgarians under the command of John Alexander. Kantakouzenos states that Manuel Asan was the head of the *vasilike taxis*, and that six soldiers of this unit were killed on the battlefield.[43] Bearing in mind that Manuel Asan was a close relative of Kantakouzenos and of the emperor and that Margarites was one of the new men Theodore II promoted to high office, it is reasonable to conclude that the *vasilikon allagion* or *taxis* (Akropolites uses both terms interchangeably) or *tagma* (Kantakouzenos uses the terms *tagma* and *taxis*) was the imperial retinue composed of members of the ruling elite who were very close to the emperor.[44] We know nothing about the size of this unit. Nor do we know whether it consisted only of officers and lacked ordinary soldiers.[45]

It is not safe to conclude that this unit was expected to participate in the actual fighting. Akropolites does not specify whether the *allagion* of Theodore II played an active role on the battlefield. Similarly, Kantakouzenos' account implies that the imperial *tagma* did not fight against the Ottomans in Pelekanos-Philokrene (1329). However, at

---

41 Akropolites, I, 121–122.

42 Kantakouzenos, I, 353.

43 Kantakouzenos, I, 466–467.

44 Manuel Asan was the son of Andronikos Asan, and grandson of John III Asan, king of Bulgaria and of a sister of Andronikos II. He was a cousin of Andronikos III and a brother-in-law of John Kantakouzenos. He was married to a daughter of Theodore Synadenos: *PLP*, 1506; Nicol, *The Family of Kantakouzenos*, 104.

45 Haldon states that the imperial *taxis* was loosely the group of officials accompanying the emperor: Constantine VII, Porphyrogennitos, *Three Treatises on Imperial Military Expeditions*, J. Haldon (ed.), (Vienna, 1990), 164.

the battle of Rosokastro (1332) six soldiers of this unit were killed. It is helpful to compare the information provided by late Byzantine authors with available information from the middle Byzantine period. Describing the famous battle of Mantzikert, where in 1071 the Byzantines were defeated by the Seljuks, the eleventh-century historian Michael Attaleiates, states that the Byzantine army was put in such a difficult position that,

> The unit which was close to the emperor, the commonly called *allagion*, had to fight. In the past, the other troops used to seize the victory and the other units close to the emperor used to stay away from the fighting.[46]

The battles of Mantzikert and of Rosokastro, despite the different political context and the three centuries that separate them, had something in common: the Byzantine forces were put in a difficult position and were defeated. Possibly, Kantakouzenos mentions the death of the six cavalrymen because it was something unusual. Moreover, Akropolites' statement that at the end of the campaign of 1255, Theodore II appointed Constantine Margarites alongside Manuel Laskaris as head of the garrison of Didymoteichon, giving them a sufficient number of troops, raises questions about the internal organisation of the imperial *allagion*.[47] This example indicates that the head of the *vasilikon allagion* could at the same time be head of a city or province commanding local troops. Another possibility is that the head of the *imperial allagion* would take over this command only during campaigns.

The *vasilika allagia*, which were mobile units established in the provinces, and were different from the imperial escort appear exclusively in Asia Minor. Two soldiers of the *vassilikon allagion,* Manuel Kaloeidas and Constantine Phrangopoulos appear in a document concerning a land dispute in Smyrna in 1293. Both of them are said to be strangers to the place. These soldiers of the *vasilikon allagion* seem to belong to imperial or central troops, and not to the local garrisons. Considering, however, that Andronikos II was in Asia Minor from 1290 until 1293, it could also be argued that they were soldiers of the imperial military

46 Attaleiates, Michael, *Miguel Ataliates Historia,* I. Pérez Martin (ed.), (Madrid, 2002), 112.

47 Akropolites, I, 123.

escort following the emperor.[48] Since nothing more is known about them, both answers to the question of the identity of these soldiers seem equally possible.

The *vasilika allagia* of Asia Minor are mentioned by Pachymeres. In one of his statements concerning the inability of the Byzantine army in Asia Minor to resist the Turkish advance at the end of the thirteenth and early fourteenth centuries he states that soldiers of the *vasilika allagia* migrated from the east because of the inroads of the Turks.[49] The sense of this statement is that these *vasilika allagia* were Byzantine troops stationed and settled in Asia Minor and were not members of the imperial escort. If this is correct, then the question is what were the differences between these *vasilika allagia* and the Asian *allagia* Pachymeres mentions, describing the composition of the army John Palaiologos in the1260s?[50] A possible answer is that Pachymeres did not make any distinction between *vasilika allagia* and *allagia*. In other words, the various *allagia* campaigning under the orders of John Palaiologos, were the *vasilika allagia* that later fled to the west. Another possible interpretation is that the *vasilika allagia* were parts of the units Pachymeres mentions with reference to John Palaiologos' army. In this context, the *allagia* is a general term, which includes all the *pronoia* soldiers established in the provinces regardless of status, wealth and equipment and the soldiers of the *vasilika allagia* were troops of higher status and therefore better equipped than the rest of the soldiers of the *allagia*.

The *megala allagia* are well documented in the sources and a good number of individual soldiers belonging to them are known. Contrary to the *vasilika allagia*, the *megala allagia* are found exclusively in the European provinces of the empire. Three *megala allagia* are known by name. The *Thessalonikaion mega allagion*, the *Serriotikon mega allagion*, and the *Vizyeteikon mega allagion*. In addition, describing the functions of the *stratopedarches ton monokavallon*, Pseudo-Kodinos states that the army was divided into *tagmata* (it should not be forgotten that *tagmata* and *allagia* are the same thing) stationed in the

[48] MM, IV, 179; Bartusis, "Megala Allagia," 185; H. Ahrweiler, "L' histoire et la géographie de la région de Smyrne entre les deux occupations Turques (1081–1317) particulièrement au XIIIe siècle," *TM* 1 (1965), 158, 161.

[49] Pachymeres, IV, 447.

[50] Pachymeres, II, 403.

provinces. Some of the soldiers were called the *megaloallagitai* (soldiers belonging to the *mega allagion*), some *trikavalloi* (soldiers with three mounts), some the *dikavalloi* (soldiers with two mounts) and others the *monokavalloi* (soldiers with one mount).[51]

The structure and role of the *megala allagia* and their differences with the *vasilika allagia* and *allagia* have attracted the attention of modern scholars and different, opposing views have been put forward. Angold, concludes that the *allagia*, were recruited from the European and Anatolian provinces and that the *megaloallagitai* were troops from the field army stationed in the cities of the European provinces of the empire.[52] Bartusis, concludes that, once Asia Minor was lost, the *vasilika allagia* disappear from the sources. He correctly observes that the *megala allagia* appear for the first time during the reign of Andronikos II only in Europe. He adds that,

> The establishment of the *megala allagia* can be dated to sometime after the beginning of the Nicaean state's re-conquest of Europe and can be seen as an effort to reconstitute a provincial administrative apparatus composed of 'thematic' regions for military as well as, fiscal purposes.[53]

By analysing the value of *pronoiai* a number of soldiers received, Oikonomides concludes that the *megaloallagitai* were more heavily equipped than the rest of the *pronoia* soldiers of the *allagia* since they were receiving substantially larger *pronoia* grants.[54]

Another issue is whether the soldiers of the *megala allagia* were linked to the town indicated by their name. Oikonomides states that, bearing in mind the large amount of the income a good part of the known soldiers of the *megala allagia* of Thessalonica and Serres received it is indicated that they were well-to-do soldiers, who could afford to participate in the campaigns bringing with them several mounts and being accompanied by one or two servants. Therefore, according to Oikonomides, the *mega allagion* from Thessalonica was the cavalry corps of Thessalonica and the *pronoia* soldiers had the obligation to serve the emperor both in their place of origin and in long-distance campaigns.[55] However, he points out that this argu-

---

51 Pseudo-Kodinos, 187.

52 M. Angold, *A Byzantine Government in Exile. Government and Society under the Laskarids of Nicaea 1204–1261* (Oxford, 1974), 189.

53 Bartusis, *The Late Byzantine Army,* 195–196.

54 Oikonomides, "À propos des armées," 353–354.

55 Oikonomides ,"À propos des armées," 353–354; idem, "The properties of the

ment relies heavily on the reliability of Pseudo-Kodinos' information, which divides the soldiers in the provinces according to the number of mounts; *monokavalloi, dikavalloi, trikavalloi, megaloallagitai.*[56] This information implies that the *megaloallagitai* could afford more than three mounts.

Bartusis argues that, the fact that some of the known *megaloallagitai* were well-to-do and cavalry soldiers, does not mean that all were wealthy cavalry soldiers. He also argues that soldiers belonging to the *megala allagia*, were serving not only the city named after the *allagion* but also the wider district and performed garrison duties as well and therefore, they could not be exclusively wealthy cavalrymen. Instead, they should have been archers and infantrymen, suited for defensive military operations of this type. However, this argument is based on the single surviving reference to the *mega allagion* of Vizye, which is an inscription on a marble block embedded in a wall of a church in a village 160 km away from Vizye. Bartusis admits the possibility that this inscription might lead to misleading conclusions.[57] Schreiner presents a different view, arguing that the *mega allagion* of Thessalonica was the unit of the imperial guard stationed in Thessalonica.[58]

These different interpretations of the role of the *megala allagia* in the late Byzantine army reflect the lack of sufficient information. Nevertheless, Oikonomides' statement that the *megaloallagitai* of Thessalonica were the wealthier pronoiars who formed the cavalry corps of Thessalonica is the most plausible interpretation. Moreover, his argument is supported also by Kantakouzenos' account, a source that Oikonomides did not use. Kantakouzenos makes a distinction among the *pronoia* holders who formed the *allagia.* In 1341, he planned to lead a campaign against the Albanians. He relates that he was planning to lead the mercenaries and 'the more powerful of those who derive their incomes from the villages.' He would spend the winter with them in the west.[59] The more powerful (*dynatoteroi*) of those *pronoia* holders would have been the wealthier soldiers, who would be able to support themselves for a longer period of time, rather than

Devlitzenoi," 177; idem, "Notes sur un praktikon de pronoiaire (juin 1323)," *TM* 5 (1973), 338.

[56] Pseudo-Kodinos, 187.

[57] Bartusis "Megala Allagia," 189–193.

[58] Schreiner, "Zwei praktika," 39.

[59] Kantakouzenos, II, 81–82.

the rest of the pronoiars, when on campaign. It is likely that these soldiers were the *megaloallagitai*, while the *trikavalloi, monokavalloi* and *dikavalloi* were three other categories of pronoiars. Furthermore, the sources show that the size of the *pronoiai* of the known *megaloallagitai* is considerably larger than the size of the other known military *pronoiai.*[60] In addition, it seems that some of the holders of large military *pronoiai* were seen as people of higher social status. The pronoiars who in 1344 donated 600 *modioi* of land to the monastery of Docheiariou are called *archontes* (lords) *stratiotai.*[61] The *megaloallagitai* therefore appear to have been wealthy soldiers. It is possibly that they did not rely exclusively on the revenues received through the *pronoiai* allocated by the state but that they expanded their personal possessions. Their actual revenues could have been substantially larger than the income they received from their *pronoiai.*[62] They were allowed to increase the value of their property and due to the government's weakness, some of them were in charge of maintaining the security and defence of the areas the incomes of which they received.[63]

Furthermore, although chronologically the *megala allagia* appear in the sources after the *vasilika allagia*, it is doubtful whether this implies any significant administrative change. Instead, it seems that it reflects mainly or merely a change of terminology. The nature of the sources should be taken into account. It is an oversimplification to argue that the *megala allagia*, which are attested in the documentary sources, and exclusively in the archives of Athos, are definitely different from the *vasi-*

[60] This is indicated by a number of examples. In 1321, Michael Saventzes and Nikolaos Maroules held *pronoia* of seventy and seventy two *hyperpyra* respectively: *Actes de Xénophon*, D. Papachrysanthou (ed.), (Paris, 1986), 137–144. In 1323, Basil Verilas held a *pronoia* of eighty *hyperpyra*: Schreiner, "Zwei Praktika," 34–36. These *pronoiai* are much higher than the ones given below. Martinos and John Panaretos *pronoiai* amounted to thirty *hyperpyra*, and the *pronoia* of Kassandrenos and later of Lependrenos, was forty *hyperpyra: Les archives de Saint-Hean Prodrome sur le mont Ménécée*, A. Guillou (ed.), (Paris, 1955), 49, 69–72; *Actes de Chilandar*, 115–118. For rather small *pronoia* see Bartusis, *The Late Byzantine Army*, 174–175. The Klazomenitai in 1342 received a *pronoia* of twelve *hyperpyra* and probably their social status was not noticeably higher than a group of peasants. However, the *pronoiai* of the above mentioned *megaloallagitai*, Saventzes and Maroules were much smaller than the *pronoiai* held by very wealthy individuals, which could reach four hundred *hyperpyra*. See Oikonomides,"À propos des armées," 353–354.

[61] *Actes de Docheiariou*, 164.

[62] Bartusis, *The Late Byzantine Army*, 172.

[63] See P. Magdalino, "An Unpublished *pronoia* Grant from the Second half of the Fourteenth Century," *ZRVI* 18 (1978), 155–163.

lika allagia mentioned by Pachymeres, bearing in mind the differences in the language used by these two different kinds of sources. Another factor, which indicates that the *megala allagia* and *vasilika allagia* were either similar or the same thing, is that individual soldiers of the *megala allagia*, Demetrios Devlitzenos, and Manuel Doukas Adrianos, soldiers of the *mega allagion* of Thessalonica in 1311, are called *vasilikoi stratiotai* (imperial soldiers), like the above-mentioned soldiers of the *vasilika allagia*, Kaloeidas and Phrangopoulos.[64] This statement indicates that *megala* and *vasilika allagia* were interchangeable.

Therefore, the troops stationed in the provinces and were made up of *pronoia* holders are the *allagia*, which were the mobile units of the late Byzantine army and were not parts of the local garrisons. Bearing in mind that the *pronoiai* were not of equal size it is reasonable to conclude that the *allagia* troops were divided according to their status and wealth. In this context, probably the *megaloallagitai* were the wealthier Byzantine heavy cavalry forces. However, there is no evidence to suggest that the *megala allagia* fought in separate units from the rest of the soldiers forming the *allagia*. This can only be a speculation based on the fact that the *megaloallagitai* were wealthier soldiers probably with better armour and equipment. Moreover, the available information does not exclude the possibility that less wealthy *pronoia* soldiers served in the *allagia* as infantry troops. Although Pseudo-Kodinos separates the *tagmata* soldiers according to their mounts and Kantakouzenos specifies that, only the wealthy cavalrymen (*stratiotai*) and Latin mercenaries are soldiers worthy of respect, they do not prove the absence of foot soldiers in the *allagia*. The above discussion shows that the exact composition and internal organisation of the armies, which the late Byzantine emperors and their generals took on campaign is not specified in the sources except in a general way. The sources use the general terms *tagmata, taxeis, stratia, stratevmata, stratopeda* without specifying differences and quite often the same unit can be identified as *taxis, tagma, allagion*.

Despite the insufficient information concerning their internal organisation, the sources show that, the troops stationed in the provinces made up a substantial part of the late Byzantine campaigning

[64] Oikonomides, "The properties of the Devlitzenoi," 177; *Actes de Docheiariou*, 119.

armies. As has been stated in another context, describing the troops led by John Palaiologos in 1262, Pachymeres mentions the *allagia* from Paphlagonia, Mesothynia, Neokastra, Mylasa, Thrace, Macedonia. He adds that the Mesothynian *allagion* was a large one and that the largest was the Paphlagonian.[65] The *allagia* mentioned by Pachymeres can be found in incidental references of Akropolites concerning the Nicaean armies, without, however, being called *allagia*; therefore, they were probably established before the recovery of Constantinople.[66] Moreover, Pachymeres reports that in 1262 Michael VIII sent to Monemvasia all the Byzantines from Magedon (Neokastra in Asia Minor) and all the Turks, meaning the Turkish mercenaries, while the *Chronicle of the Morea* mentions that Michael VIII sent to Morea an army composed of Turks, Cumans and Byzantines from Nicaea.[67] Gregoras states that the forces of John Palaiologos in his campaign against the *sevastokrator* John Angelos, the ruler of Thessaly in 1273, included Paphlagonian cavalry and troops from Bithynia.[68]

Kantakouzenos, throughout his account, indicates that the core of the army was made up of soldiers based on provinces and cities. For instance, on the eve of the civil war of 1321–1328 he states that the armies of the western parts of the empire were very well trained in battles, while Andronikos III contemplating his moves stated that,

> The armies established in the Thracian cities up to Christoupolis have already joined our side. However, there are many proud cities beyond Christoupolis, as far as Akarnania and Dalmatia, which are subjects of the Romans and have considerable military forces and good generals, which have not joined us.[69]

---

[65] Pachymeres, II, 403.

[66] Akropolites (I, 139, 146–148) reports that in 1256 Theodore II sent a small army of Paphlagonians alongside 300 Cumans, under the command of Michael Laskaris, to perform garrison duties in Thessalonica. The following year, the same Paphlagonians fought alongside Michael Palaiologos against Michael II Angelos of Epiros. In this instance, they were not part of a campaigning army, since according to Akropolites' account, this army was too small to fight pitched-battles. However, Akropolites points out that when, in 1257, Michael II Angelos counterattacked the Nicaeans in Macedonia, these Paphlagonians were the best troops available to the Nicaeans. He adds that this army was composed of fifty men. For the terms Akropolites uses to refer to the army see Macrides, *The History*, 98.

[67] Pachymeres, I, 273; *Chronicle of the Morea*, vv. 4553–4557.

[68] Gregoras, I, 111.

[69] Kantakouzenos, I, 101.

In 1328, Andronikos III had summoned the armies from the Macedonian and Thracian cities as well as from Constantinople.[70] The army Andronikos III led in Bithynia in 1329 was composed of the armies (*stratiai*) of Constantinople, Didymoteichon, Adrianople and Thrace.[71] It seems logical to conclude that these armies were the *allagia* stationed in large and important cities. It should not be forgotten that Kantakouzenos does not use contemporary technical military terminology and he never mentions the term *allagia*. Instead, he prefers the term *stratia*.

The Nicaean frontier troops, the Proselantai *and the* Thelematarioi

Grants of land were also used for the financing of the troops stationed in the eastern frontier during the Nicaean period, the *proselantai* and the *thelematarioi*. The limited available sources show that these groups of soldiers received small grants of lands, which often were not the only means of their financing. These soldiers were less prestigious than the wealthy pronoiars. The role of the Nicaean frontier guards is known through the preface of Pachymeres' account. He writes that,

> Not to ignore those living towards the mountains, who were ready to migrate, if the enemy should attack anywhere, since they [the emperors] did not have anything to persuade them to remain, if this thing should happen [enemy attack], and to resist the enemy with bravery and courage, they granted tax exemptions to all, and *pronoiai* to the most notable, and those with a more daring spirit were granted imperial letters. In time, they increased their properties and accumulated much wealth. The more they had in abundance the necessities of life, the more courageous they grew against the enemies and they enjoyed many of the goods of those on the other side of the frontier by laying ambushes at night and by daily ravaging the land of the enemy, gathering much booty; because of these, it happened that their attacks against the enemy helped those living in the interior and hoping to live comfortably in their lands and occupy themselves with their domestic affairs, . . . and the people of the frontiers enjoyed not only exemptions and *pronoiai* but also received daily proofs of friendship from the emperors and everything indicated their wealth; and because of this they became more ardent guards and invaded and confronted those who ravage our lands with greater bravery.[72]

---

[70] Kantakouzenos, I, 326.
[71] Kantakouzenos, I, 342.
[72] Pachymeres, I, 29–31.

It is possible that Akropolites' statement that Philadelphia was a large city located on the borders with the Turks, whose residents are very competent soldiers and are in constant warfare with the Turks, refers to the same troops.[73] According to Pachymeres, the frontier soldiers were granted tax exemptions, financial grants and the most notable of them received *pronoiai*. Therefore, they received their remuneration through the combination of different systems. These frontier troops seem to have been the first line of defence against the Turcoman raids. They were even able to raid on the enemy territory for their own enrichment and to damage the morale and the resources of their opponents. This practice is reminiscent of the instructions given by a late tenth-century military treatise attributed to the general Nikephoros Ouranos. As Ouranos writes,

> The generals on the borders ought to conduct continuous raids, so that when a favourable time comes and the army comes out to besiege this fortress it will find it well below strength in manpower and in shortage of provisions.[74]

Furthermore, the establishment of local provincial troops in the frontier reflects the main strategic aim of the Nicaean rulers with regard to Asia Minor, which was the maintenance of peace and stability in the eastern frontier. This would enable them to concentrate all their military efforts and resources on the recovery of Constantinople and the European provinces of the old Byzantine empire, without having to fight on two fronts.

It is worth noting that, the Nicaean emperors were not the only thirteenth century rulers who followed this frontier policy. In the thirteenth-century Christian-Muslim frontier in Spain, the Spanish rulers established settlements in easily defended places and natural strongholds to secure the occupation of the frontier. The result was the development of a society of frontier warriors, who carried out raids on the other side of the frontier, collecting considerable booty.[75]

Therefore, the strategic role of these frontier guards was to secure the continuous occupation of lands close to the frontier. By defending

[73] Akropolites, I, 95.

[74] Mc Geer, *Sowing the Dragon's Teeth*, 152–154, whose translation this is.

[75] M.G. Jimenez, "Frontier and Settlement in the Kingdom of Castile (1085–1350)," in R. Bartlett - A. Mackay (ed.), *Medieval Frontier Societies* (Oxford, 1989), 54–55; J.F. Powers, *A Society Organized for War. The Iberian Municipal Militias in the Central Middle Ages. 1000–1284* (Los Angeles/London, 1986), 40, 64.

those territories they would act as a buffer between the Turkish raiders and the rich valleys of western Asia Minor, the heart of the Nicaean Empire. To connect these troops with the land they had to defend, would make them more determined defenders and would ensure that the frontier regions were not deserted. Moreover, the local army of non-professional soldiers responds more efficiently to the needs of constant small-scale warfare, like that between the Nicaeans and the Turks, than does an army of professionals or mercenaries. The latter were expensive for the government, which had to pay not only for their subsistence but also their service, and it was not easy to employ them for long. Pachymeres and Gregoras consider the establishment of the frontier troops in Asia Minor as a successful development. They also point out that the abolition of this system and the deployment of these troops in campaigns in Europe, was one of the main causes for the collapse of the defence in Asia Minor in the late thirteenth and early fourteenth centuries.[76] The view that the frontier is better defended, if it is densely populated, can be found also in the last years of Byzantium. In a letter dated between 1444 and 1446 which Bessarion sent to the despot of the Morea, Constantine Palaiologos—the future emperor Constantine XI—, he suggested that the wall of Hexamilion, which was built to protect the Peloponnese from the Ottomans, could be better defended, if the despots were to build a city in the same area. As he states, walls without guards cannot do anything good and guards without a city close to them are not able to perform their duties, nor do they feel secure.[77]

According to Pachymeres the *Thelematarioi* were the inhabitants of Chryseia in the vicinity of Constantinople during the period of the Latin empire and they changed sides inclining towards the Byzantines or the Latins, at will. For this reason they were called *Thelematarioi*. The Latins did not banish them, fearing the economic consequences of the desolation of the area.[78] The *Thelematarioi* were farmers, whose products were available to the Latins in Constantinople. Michael VIII granted them land, which they could bequeath to their children, as a reward for their co-operation in the recovery of Constantinople by

[76] Pachymeres, I, 31–35; Gregoras, I, 138.
[77] *PP*, IV, 35.
[78] Pachymeres, I, 157. For the term *Thelematarioi* see Geanakoplos, *Emperor Michael*, 95–96.

the Byzantines in 1261.[79] The sources provide insufficient information concerning their role in late Byzantine military operations. In 1305, they appear as a group only in the battle of Apros against the Catalans, probably as infantry troops. The chronological distance between the battle of Apros and the recovery of Constantinople indicates that those who fought at Apros were the sons and grandsons of the original *Thelematarioi*.[80]

Pachymeres states that the *Proselantai*, who were the rowers of the fleet established by Michael VIII were recruited from every part of the Empire and settled in the coastal areas of the empire.[81] They participated in the naval campaigns of the Byzantine fleet in the 1260s and 1270s and they are found in the sources long after the reduction of the fleet carried out by Andronikos II in 1285.[82] They do not appear as rowers of the fleet in the fourteenth century. This may be because Kantakouzenos and Gregoras avoid using contemporary technical military terms.[83] Very little is known about their organisation. It is not known whether they were mercenaries like the sailors of the fleet, the *Gasmouloi*. Nevertheless, a number of documents show that at least some *Proselantai* had received grants of land.[84]

The case of the despotate *of the Morea*

With regard to the financing of soldiers through grants of land, it is useful to discuss briefly the writings of the philosopher George Gemistos Plethon. In his *Address to Manuel II on Affairs in the Peloponnese*, written in 1417 or 1418, he implies that the bulk of the army in the fifteenth-century Morea was made up of farmers.[85] He states that, most of the Peloponnesians, who are called up in the army, are farmers, who,

---

[79] Pachymeres, I, 221.

[80] Pachymeres, IV, 599–601; Bartusis, "Smallholding Soldiers," 14–15.

[81] Pachymeres, I, 223, II, 403.

[82] For the documentary evidence on the *proselantai* see Bartusis, "Smallholding Soldiers," 17–20.

[83] K.-P. Matschke, "Johannes Kantakuzenos, Alexios Apokaukos und die byzantinische Flotte in der Burgerkriegsperiode, 1340–1355," in *Actes du XIVe Congrès international des études byzantines* (Bucharest, 1971), 204.

[84] Bartusis, "Smallholding Soldiers," 17–19; Oikonomides, "À propos des armées," 357–358.

[85] *PP*, III, 246–265.

since they do not earn much, do not stay for long in the campaigns. Instead, they feel the need to return to their homes to take care of the affairs of their households. Gemistos suggested the soldiers be set apart from the taxpayers. He also suggested to the despot Theodore II Palaiologos (1407–1443) that the agricultural workers be grouped in pairs, each alternating military service and labour. His account suggests that the farmers were recruited on a campaign by campaign basis, and that they did not receive any compensation for their service, apart from the prospect of booty. Therefore, from Plethon's point of view the proper defence of the Peloponnese should rely on soldiers who would not need to worry about their material needs. Gemistos' proposals did not materialise. Possibly, the Byzantine authorities did not have the resources to carry out such reforms. Moreover, Gemistos' reforms implied social changes that probably the Byzantine rulers were not willing to support in view of the opposition they could arouse from the local elite which was constantly prone to revolt.

Plethon's ideas have been the subject of debate. There are scholars who argue that his proposals offered a concrete solution to the problems of the defence of the Peloponnese. Others maintain that Plethon expresses merely his desire to present a classical Greek solution.[86] It has also been suggested that Plethon was influenced less by ancient authors and more by the structure of his contemporary Ottoman army.[87] Plethon had spent time in Ottoman Bursa (Prusa) where he received part of his education and was familiar with the military organisation of the Ottoman empire. In his writings he remarks that, the state organisation of the Turks (the 'barbarians'), favoured military

[86] *PP*, III, 251. For an analysis of Plethon's proposals see C.M. Woodhouse, *George Plethon Gemistos: The Last of the Hellenes* (Oxford, 1986), 100–106; Bartusis (*The Late Byzantine Army*, 217–221) states that, 'Gemistos' account is a fabrication, which derived from his desire to present a classical Greek situation.' Blum considers Gemistos' proposals as a realistic solution to the problems of the Morea: W. Blum, *Georgios Gemistos Plethon. Politik, Philosophie und Rhetorik im spätbyzantinischen Reich* (Stuttgart, 1988), 36–44, 76–80. Nicol relates that the materialisation of Plethon's proposals required the presence of a strong central authority which did not exist in the Morea: D. Nicol, *The Immortal Emperor* (Cambridge, 1992), 22–23; Matschke and Tinnefeld (*Die Gesellschaft*, 361–362) argue that Gemistos' proposals were not mere imitation of Plato's thought. They reflected the political realities of the fifteenth-century Peloponnese. They could not materialise because of the absence of a strong central government which could undertake such reforms.

[87] Necipoğlu, *Between the Ottomans and the Latins*, 275.

success and expansion.[88] Similar proposals were put forward by Plethon's student, Bessarion, who, like his teacher, might have been influenced by Ottoman military practices. In a letter to Constantine Palaiologos written sometime between 1444 and 1446 Bessarion comments that the Peloponnesians lacked military training and were stripped of weapons due to the cruelty of their lords and the harsh exactions that are imposed upon them. In order to improve the effectiveness of the Peloponnesian army Bessarion suggested the division of the population into two groups one of which will be occupied strictly with military matters (*to stratiotikon*) while the other will be engaged exclusively in agriculture (*to georgikon*).[89]

Apart from the writings of Plethon and Bessarion, the available information concerning the military organisation of Byzantine Morea is insufficient to draw conclusions. Describing the preparation of the despot Constantine Palaiologos for the siege of Patras in 1429, Sphrantzes, who had participated in this operation, relates that the heads (*kephalades*) in the area around Androusa received written orders to assemble in Kastri, where the despot set up camp, with their arms and those who are under their command.[90] Relying on this evidence, Zakythinos concludes that, the army in the Peloponnese relied on a kind of vassalage.[91] Bartusis disagrees. He also rightly doubts the fighting abilities of the Moreot army, basing his argument on the fact that neither Constantinople nor any other city was ever aided by reinforcements from the Peloponnese.[92] For instance, in 1448 Constantine XI did not bring any troops from the Peloponnese to Constantinople to secure his accession.[93] A contemporary source, the *Cronicle of the Toccos* seems to justify this view stating that Byzantine troops in the fifteenth-century Morea were often receiving reinforcements from Constantinople. For instance, the anonymous author states that in 1417/1418 Manuel II brought weapons, soldiers and ships from Constantinople in order to fight against the Navarrese and Centurione

---

[88] *PP*, IV, 118. For Plethon's sojourn in Bursa see Woodhouse, *George Plethon Gemistos*, 26–27.
[89] *PP*, IV, 35.
[90] Sphrantzes, 42.
[91] Zakythinos, *Despotat*, II, 137, 138.
[92] Bartusis, *The Late Byzantine Army*, 119.
[93] M. Klopf, "The Army in Constantinople at the Accession of Constantine XI," *B* 40 (1970), 389.

Zaccaria, the Latin prince of Achaia (1402–1432).[94] Although he says nothing about the nature of the military conflicts in the Peloponnese, Doukas seems to confirm the testimony of the anonymous chronicler. He relates that Manuel II sailed to the Peloponnese and achieved significant successes against the Navarrese and the prince of Achaia.[95]

[94] *Cronaca dei* Tocco, vv. 3531–3540.
[95] Doukas, 139.

CHAPTER FOUR

MERCENARIES

The use of mercenaries

The limited financial resources of the Byzantine state did not deter late Byzantine rulers from employing large numbers of mercenaries. Despite the high cost of their maintenance, their employment served important political and military purposes. Throughout his account John Kantakouzenos, an emperor and soldier himself, emphasises the need to employ mobile troops who could be almost constantly under arms. He writes that, in 1343 his wife, Eirene Asanina, told him not to wait for a military force to be assembled for him. Instead, Kantakouzenos should receive as many German mercenaries from the Serbian ruler as possible because, 'since they served for pay, they were always ready for campaigns.'[1] With reference to the arrears the state owed to its troops in 1327, Kantakouzenos commented that it is unjust, brutal and inhumane for mercenaries not to receive their payment, since they are not allowed to be involved either in commerce or in agriculture or in any other occupation, for they have to be always ready to campaign and should not have any hindrance that would force them to stay at home and prevent them from campaigning.[2] In another similar example, in the winter of 1341, Kantakouzenos had planned to undertake a campaign against the Albanians in Epiros leading only the mercenaries and the wealthier of the pronoiars, giving time to the rest of the army to prepare for the following spring campaign against the Latin principality of Achaia in the Peloponnese.[3]

Foreign mercenary soldiers covered the tactical inefficiencies of the late Byzantine army and brought Byzantine warfare closer to the world around it. Western European mercenaries provided the Byzantines with heavy cavalry soldiers and tactics, and with highly disciplined infantry, such as the Almugavars, who were part of the Catalan Grand

---

[1] Kantakouzenos, II, 354.
[2] Kantakouzenos, I, 238.
[3] Kantakouzenos, II, 81–82.

Company. The Turks and the Cumans were light cavalry archers, whom the Byzantines used very effectively in 1259 against German and French knights in the battle of Pelagonia, in the 1260s against the Franks in the Peloponnese, in 1281 against the troops of Charles I of Anjou in Bellagrada and probably in the 1270s and 1280s in the operations against Angevin possessions in Epiros and Albania. Mercenaries could provide the Byzantines with specialised professional soldiers who possessed advanced knowledge of military technology. Describing the siege of Philippoupolis (modern Plovdiv) in 1323 by the troops of Andronikos III, Kantakouzenos relates that a German soldier who was trained in the construction of siege engines suggested the construction of a wheeled siege tower which was moved by 100 soldiers.[4] Advanced technological knowledge was provided by foreign specialists, like the engineer John Grant, who in the siege of Constantinople in 1453 supervised the repairs on the city's walls and was in charge of preventing the Ottomans from successfully mining the Byzantine fortifications. In his account of the fall of Constantinople the Latin bishop of Chios, Leonard, calls him a gifted soldier who was skilled in every aspect of the art of war.[5]

The most important political motive behind the employment of mercenaries was that they lacked strong ties with the localities in which they served. The effect of the local affiliation of the native troops is apparent in Kantakouzenos' account. In 1342, during the civil war of 1341–1347, Kantakouzenos reached Thessalonica and set up his camp in the Galikos river, where he received Theodore Synadenos with 1000 cavalry soldiers, who had been expelled from Thessalonica. Kantakouzenos held a war council and suggested that the army not attempt any attack against Thessalonica. Instead, they would leave only a small force in the fortress of Gynaikokastro to carry out raids on Thessalonica. The bulk of the army would march on Veroia and Edessa, which was besieged by the Serbians. However, the soldiers from Constantinople and Thessalonica strongly opposed this view. Kantakouzenos comments that they were full of fear and agony because of the unstable situation in their home cities. This made them think that the march on Epiros would be long and tortuous. According

[4] Kantakouzenos, I, 173–174.

[5] A. Pertusi, *La Caduta di Costantinopoli*, 2 vols. (Rome, 1976), I, 135, 149; J.-R. Melville Jones, *The Siege of Constantinople 1453. Seven Contemporary Accounts* (Amsterdam, 1973), 17, 27; Chalkokondyles, II, 152.

to Kantakouzenos, the soldiers from Thessalonica and Constantinople found many justifications to put off the march and a number of them deserted him. Eventually, the whole operation was cancelled because the army could not cross the Axios river, which was flooded, and because of the dissension among the soldiers.[6]

The recruitment of mercenaries in late Byzantium should be examined in the context of contemporary military developments in the wider geographical context. Thirteenth and fourteenth-century Europe saw the increasing use of mercenaries which was the result of a combination of economic and political developments with military needs. In Italy, which geographically was close to Byzantium, the growing wealth of cities enabled them to hire large numbers of mercenaries. In addition, the increasing use of weapons which required long practice to use effectively, such as the crossbow, the development of heavy armour which was physically demanding, the use of artillery and the impact of gunpowder on warfare encouraged specialisation.[7] Moreover, the enemies and neighbours of Byzantium relied heavily on mercenaries. The Frankish forces in the battle of Pelagonia included German cavalry.[8] The armies of Italian rulers established in Frankish Greece, such as the Acciaiuiolis, who were members of a wealthy Florentine family of bankers, and the Toccos, were predominately mercenary forces. The Toccos are said to have employed Serbians, Byzantine Greeks, Latins and Albanians.[9] In the late 1370s and 1380s the companies of Navarrese and Gascon soldiers dominated warfare in Frankish Greece and Albania. They were hired by the Hospitallers who had received the principality of Achaia from Joanna of Naples, the Angevin claimant to the throne of the kingdom of Albania whose husband, Louis of Evreux, Count of Beaumunt-le-Roger, had employed the Navarrese companies of soldiers to assist them in their war against the Albanians.[10] The army of Nerio Acciaiuioli, which in 1379 captured Thebes and Megara from the Catalans of Athens, included Navarrese mercenaries.[11]

---

[6] Kantakouzenos, II, 237–243.

[7] See M. Mallett, *Mercenaries and their Masters*, (London, 1974), 18–20.

[8] Akropolites I, 168.

[9] *Cronaca dei Tocco*, vv. 134–137, 201–202, 549, 826, 985, 1071.

[10] R. Loenertz, "Hospitaliers et Navarrais en Grèce 1376–1383. Régestes et documents," *OCP* 22 (1956), 331–332; G.T. Dennis, "The Capture of Thebes by the Navarrese (6 March 1378) and other Chronological Notes," *OCP* 26 (1960), 45.

[11] A.R. y Lluch (ed,), *Diplomatari de l'Orient Català (1301–1409)* (Barcelona, 1947 repr. 2001), 537–538; *MP*, 33.

Shortly afterwards they were in the service of Jacques of Baux, who was a claimant to the principality of Achaia and when he died in 1383 they claimed it for themselves.[12] The Bulgarians employed Cuman and Tatars, the Serbians Germans, Catalans, and Cumans. Furthermore, the Seljuk sultans of Rum employed Turcoman, Byzantine and Latin mercenaries.[13] A characteristic example showing the 'international' and mercenary character of warfare in the Balkans was the Bulgarian army defending Philippoupolis against the besieging Byzantine army, which included an unknown number of Western European mercenaries. The city was defended by 1000 Alan cavalry and 2000 light infantry, whose identity is not specified by Kantakouzenos. Their commanders were two Alan generals and a Hungarian, while the overall command was in the hands of a certain Ivan the Russian.[14]

Categories of mercenaries

The mercenaries employed by the Byzantines can be divided into two main categories: Those employed for long-term service and those hired only for a specific occasion. Most of the known cases of mercenaries hired for long-term service are members of large groups of people, who shared a common cultural background and were settled in the empire. The Cumans made up a substantial part of the Nicaean and Palaiologan armies from 1240 until the end of the thirteenth century. Modern scholarship unanimously agrees that the Cumans were smallholding soldiers.[15] However, the sources do not say anything about their maintenance. Gregoras states that John III enlisted them in the Nicaean forces and provided them with lands for habitation in Thrace, Macedonia, the Meander valley and in Phrygia.[16] He does not say that they received these lands in exchange for military service. It is likely that they served as long-term mercenaries who were paid only

[12] K. Setton, *Catalan Domination of Athens. 1311–1388* (Cambridge, 1948), 125–148; Bon, *La Morée franque*, 254–261; A. Luttrell, "Appunti sulle compagnie navarresi in Grecia," *Rivista di studi bizantini e slavi* 3 (1983), 113–127.

[13] Bombaci, "The Army of the Seljuqs of Rum," 353–363.

[14] Kantakouzenos, I, 173.

[15] Bartusis, "Smallholding Soldiers," 12; C. Asdracha, *La region des Rhodopes aux XIIIe et XIVe siècles: étude de géographie historique* (Athens, 1976), 81, 242–243; J. Langdon, *John III Ducas Vatatzes' Byzantine imperium in Anatolian Exile, 1222–54* (Los Angeles, 1978), 249–250; Angold, *Exile*, 105.

[16] Gregoras, I, 37.

when they campaigned and the lands they received were irrelevant to military service. Therefore, while the Cuman soldiers were permanent residents of the empire, they were called into arms and paid as casual mercenaries.

The strategic importance of the enlistment of the Cumans is discussed by Theodore II and Akropolites. Theodore II praises his father John III for having removed the Cumans from the western lands and led them to the east as subject people, fettering their assaults on the west.[17] Akropolites considers them 'a battle-worthy army,' which is his standard phrase for a good-sized army.[18] He adds that John III transformed them from a savage people to an obedient one, and he drew them away from Macedonia and brought them to the east.[19] Therefore, by enlisting the Cumans the Nicaeans relieved their western provinces from their attacks and reinforced their armies with a significant number of light cavalry, who served them for the following fifty years. The Cuman soldiers participated in all the military operations of the Nicaean armies after 1240. After 1261, their presence is attested in the campaigns of John Palaiologos in Epiros in the 1260s and in Thessaly in the 1270s. They appear for the last time in a campaign against Epiros in 1292.

The Cuman light cavalry archers were renowned for their military competence long before a number of them joined the Nicaean armies. They had allied with various Russian princes and Georgian kings and because of their fame Philip of Swabia, one of the candidates in the struggle for the German throne had hired a body of them in 1203.[20] In 1239, a large number of Cuman warriors joined the Hungarian army.[21] Furthermore, they made up a large part of the Bulgarian armies and a number of them joined the Latins of Constantinople.[22] In 1205, their role in the defeat of the Latins of Constantinople at the hands of the Bulgarians in the battle of Adrianople was decisive.[23]

---

[17] Theodore II, *Encomio*, 49.

[18] Macrides, *The History*, 143.

[19] Akropolites, I, 65.

[20] P. Jackson, *The Mongols and the West 1221–1410* (London, 2005), 17.

[21] Jackson, *Mongols*, 61; I. Vásáry, *Cumans and Tartars. Oriental Military in the Pre-Ottoman Balkans, 1185–1365* (Budapest, 2005), 65.

[22] *Synopsis Chronike*, 480–481; Vásáry, *Tatars and Cumans*, 66.

[23] Choniates, *Historia*, 616–617; Akropolites, I, 21–22; Gregoras, I, 15–16; *Chronicle of the Morea*, vv. 1114–1161; Villehardouin, *La conquête de Constantinople*, in A. Pauphilet (ed.), *Historiens et chroniqueurs du Moyen Age* (Paris, 1952), 169–172; de Clari, *La conquête de Constantinople*, P. Lauer (ed.), (Paris, 1974), 105–106.

Very little is known about the Albanians who were employed by the Byzantines in the Peloponnese. In the funeral oration to his brother, Manuel II, referring to the Albanian migration to the Peloponnese in 1394–1395 relates that around 10,000 Albanians left their lands and together with their children, women, property and cattle occupied the Isthmus pitching their tents there. They were settled in uninhabited lands, which they cultivated. Manuel II adds that to have the Albanians together with the forces of the Peloponnese which were not small was of the greatest assistance. Manuel II relates that their role was to reinforce the defence of the Morea and to help the despots to impose their authority over the local lords. Their presence is attested in the defences of the walls of Hexamilion in the Isthmus of Corinth and in 1395 in the victory of Demetrios Raoul against the Navarrese.[24] It is also said that, many Albanians who survived the capture of Argyrokraston by the Ottomans in 1418 were recruited as mercenaries by the Byzantines in the Morea.[25] It seems that in the late fourteenth and fifteenth centuries there was a significant demand for Albanian soldiers. Albanian mercenaries served in the wars between the Latin principalities established in western Balkans. Many of the soldiers of the army of the Toccos were Albanian mercenaries.[26] Some of the Albanians who in 1398 repopulated Argos after its destruction by the Ottomans, enlisted in the Venetian army.[27] It is also worth noting that, in the mid-fifteenth century Albanian light cavalry mercenaries were recruited by Venice and used in the warfare against the Ottomans in the 1460s in the Peloponnese.[28] There is no surviving information indicating the financing of the Albanians recruited by the Byzantine governors of the Morea. Given the limited resources of the despotate it is most likely that, like the Cumans under the Laskarids, they served as long-term mercenaries who were paid only when they campaigned.

The Gasmouloi were the product of mixed marriages between Byzantines and Latins. Pachymeres and Gregoras call them people of mixed race and excellent soldiers, whose military skills combined

---

[24] Manuel II, *Funeral Oration,* 119–125; *Kleinchroniken*, I, 245; Sphrantzes, 22. For Demetrios Raoul see *PLP*, 24116.

[25] *Cronaca dei Tocco*, vv. 3292–3303; Nicol, *The Despotate of Epiros. 1267–1479*, 189.

[26] *Cronaca dei Tocco*, v. 134–137, 201–202, 549, 826, 1071.

[27] *Kleinchroniken*, I, 245; *MP*, 200.

[28] See M. Mallett – J.R. Hale, *The Military Organization of a Renaissance State. Venice c. 1400 to 1617* (Cambridge, 1984), 119, 196.

the prudence of the Byzantines with the boldness of the Latins.[29] They manned Michael VIII's fleet and they had considerable successes in the Aegean Sea against the naval forces of the Latins who were established in the Aegean islands after the Fourth Crusade. When Andronikos II, following the advice of his counselors reduced the size of the fleet, their role in the Byzantine fleet seems to have declined.[30] However, they did not disappear. Probably, they played a significant role in the civil war of 1341–1347.[31] The sources do not specify whether the Gasmouloi were paid only during military operations or not. That they were permanent residents of the empire means that we cannot exclude the possibility that they received their payment only for the purposes of the campaigns they participated.The Cretan refugees were established in Asia Minor by Andronikos II sometime before 1295. They were to receive annual grants and they provided the army with cavalry troops.[32] They disappear from the sources after the suppression of the rebellion of Alexios Philanthropenos in Asia Minor; although they instigated it, they later opposed and captured Philanthropenos. The Alans were recruited in 1301 and the following year they were sent to Asia Minor. A part of them was sent to Bithynia under the command of Leo Mouzalon and another part was sent to campaign in Magnesia under the orders of Michael IX.[33] The above-mentioned groups of people provided the army with mercenaries, whenever they requested to do so. It seems logical to conclude that, since they were permanent residents of the empire, they also possessed lands, which provided them with their income during periods of peace. Therefore, while these groups consisted a permanent source of mercenary troops for the state, their soldiers were recruited on a casual basis.

[29] Pachymeres, I, 253; Gregoras, I, 94.

[30] Pachymeres, III, 81–83; Gregoras, I, 174–176.

[31] D. Jacoby, "Les Vénitiens naturalises dans l' Empire byzantine: un aspect de l'expansion de Venise en Romanie du XIIIe au milieu du XVe siècle," *TM* 8 (1981), 221–224; Matschke, "Byzantinische Flotte," 194–196; Bartusis (*The Late Byzantine Army*, 69–70) points to the existence of fifteenth century references to the 'service of a Gasmoulos,' which indicates that the later Gasmouloi were not mercenaries and that service as a Gasmoulos lost its distinctive ethnic basis and was not restricted only to marine duties, but included service as light armed soldiers in the army as well. Oikonomides shows that in the mid-fourteenth century some Gasmouloi were merchants who could not man the military fleet: N. Oikonomides, *Hommes d' affairs Grecs et Latins à Constantinople. XIIe–XVe siécles* (Montreal, 1979), 87.

[32] Pachymeres, III, 237; Gregoras, I, 197.

[33] Pachymeres, IV, 337; Gregoras, I, 205. For Leo Mouzalon see *PLP*, 19443.

Other groups of mercenaries committed to long-term service are those who received *pronoia* grants. During John III's reign the first Latin pronoiars, the *kavallarioi*, appear.[34] The granting of *pronoiai* to Latin mercenaries can be seen as both a means of rewarding soldiers and a means of reducing the financial cost of their maintenance.[35] Bounding also the mercenaries to the land helped the state to secure their long-term commitment. However, it is impossible to assess what percentage of Frankish mercenaries received *pronoia* grants. In another example, in the treaty of Epivatai, concluded between Andronikos II and Andronikos III in July 1322, Andronikos III stated that he had increased the cash payment of the mercenaries in the area under his jurisdiction and that he had granted each of them land.[36] By granting land to the mercenaries, Andronikos III not only increased their income, but also created a bond between the mercenaries and the land, providing them with an extra incentive to remain under his service for longer, without burdening the treasury.[37] This measure was taken in the context of the civil war of 1321–1328, when both sides were desperate to secure the loyalty of as many troops as they could, while very limited financial resources were available.

The *Tzakones* were another group of mercenaries, who received grants of land. They were Peloponnesians brought to Constantinople by Michael VIII. Pachymeres indicates that they were used for the defence of Constantinople and for the maintenance of public order. He remarks that Michael VIII had great need to settle light-armed soldiers in Constantinople and so he had many Lakones, as Pachymeres calls the *Tzakones*, arriving from the Morea. Michael VIII also bestowed on those *Tzakones* annual pay and other grants and used them inside and outside the city, since they displayed worthy behaviour in wars.[38] Therefore, the *Tzakones* were entitled to an annual wage like the mercenaries, while they were settled as natives near Constantinople.[39]

To the above groups of mercenaries who were not financed exclusively through cash payments, we should add the companies of soldiers since their members are found to possess *pronoiai*. Byzantine companies

[34] For details on *Kavallarioi* see M. Bartusis, "The Kavallarioi of Byzantium," *Speculum* 63 (1988), 343–351.
[35] Bartusis, *The Late Byzantine Army,* 28–29.
[36] Kantakouzenos, I, 164–165.
[37] Oikonomides, "À propos des armées," 359.
[38] Pachymeres, I, 253; Bartusis, "Smallholding Soldiers," 16–17.
[39] Pachymeres, I, 253.

of soldiers appear in the fourteenth century sources. The Varvarenoi was a group of collective *pronoia* soldiers, who held a grant of *pronoia* from around 1327 through the late 1340s in Chalkidike.[40] In 1341, John Vatatzes deserted Kantakouzenos and went to Constantinople leading the *taxis* of the Achyraitai.[41] In 1342 the Klazomenitai appear in Serres.[42] In a document of 1370, appears the company of the soldiers under the command of the judge of the army, Sgouros.[43]

The most important issue concerning the administration of the Byzantine companies of soldiers is whether they were really mercenaries or pronoiars since they always appear as holders of collective *pronoiai*. It could be argued that they were mercenaries, whom the state found convenient to maintain through *pronoia* grants.[44] This would be advantageous for the state. By providing them with the incomes of properties, the government did not need to pay them cash and secured their long-term commitment. However, the sources do not allow conclusions concerning the length of service of the known soldiers of the companies and the scarcity of the available information prevents us from giving a definite answer to the question whether they were mercenaries or collective *pronoia* holders who formed a company of soldiers. In conclusion, the distinction between a mercenary and a *pronoia* soldier was not always clear.

Nonetheless, the late Byzantine governments did hire companies of foreign mercenaries, the most famous case being the Catalan Grand Company. Possibly, the Byzantines employed more companies of foreign soldiers throughout the period under discussion. One of them might have been the 300 Catalans whom Kantakouzenos hired in 1352. They were part of the Catalan fleet that had arrived in Constantinople and fought alongside the Venetians against the Genoese of Galata. They decided to stay in Constantinople and Kantakouzenos relates that he knew their leader Juan Peralta. Peralta was serving in the

---

[40] *Actes de Docheiariou*, 142, 161–162, 164, 170, 234; Oikonomides, "À propos des armées," 360–363; Bartusis, *The Late Byzantine Army*, 201.

[41] Kantakouzenos, II, 180.

[42] Laiou, *Peasant Society*, 142–143; J. Haldon "Limnos, Monastic Holdings and the Byzantine State," in A. Bryer - H. Lowry (eds.), *Continuity and Change in Late Byzantine and Early Ottoman Society* (Birmingham, 1986), 184.

[43] Bartusis, *The Late Byzantine Army*, 202.

[44] For a discussion on the composition and identity of the late Byzantine companies of soldiers see Oikonomides, "À propos des armées," 360–371; Bartusis, *The Late Byzantine Army*, 200–205.

Serbian army in the 1340s when Kantakouzenos was at the court of Stefan Dušan in Serbia.[45] Therefore, it is likely that Peralta was the head of a company of soldiers. In 1329, in the battle of Pelekanos-Philokrene a Bulgarian named Sevastopoulos was leading 300 soldiers. Kantakouzenos' account indicates that Sevastopoulos was commanding his own company of soldiers.[46]

The limited available source material does not indicate whether the Byzantines employed companies of soldiers for the defence of fortifications similar to the company (*syntrophia*) of ballista handlers which was led by Michael Syrianos. This company was employed by the duke of Athens Nerio Acciaiuioli in the 1380s.[47] Moreover, the evidence provided by contemporary Venetian archives shows that the defence of Venetian fortresses in Frankish Greece and in the Aegean Sea relied on small companies of mercenary soldiers who were mostly crossbowmen.[48] In the 1270s, the Angevin forces in Albania recruited many crossbowmen, while the author of the *Chronicle of the Toccos* indicates that crossbowmen were essential for the defence of fortifications.[49] It is logical to infer that similar small companies were employed by the Byzantine state for the defence of fortresses and fortified towns. Moreover, in his account of the siege of Philippoupolis in 1323, Kantakouzenos points out the effectiveness of the crossbow in siege operations. He relates that the shots of the crossbowmen of the Byzantine army penetrated the armour of the defenders killing and injuring many of them.[50] The importance of this weapon in late Byzantium is indicated by the creation of the office of *stratopedarches ton tzangratoron* (crossbowmen).[51]

With reference to the role of the companies of soldiers, it should be mentioned that the increasing need for professional specialist soldiers

---

[45] Kantakouzenos, III, 244; For the battle of the Bosporus see M. Balard, "À propos de la bataille du Bosphore. L'expedition genoisie de Paganino Doria à Constantinople 1351–1352," *TM* 4 (1970), 431–470.

[46] Kantakouzenos, I, 354; Oikonomides, À *propos des armées*, 364.

[47] D.K. Giannakopoulos, *Δουκάτο των Αθηνών. Η κυριαρχία των Acciaiuioli* [*The Duchy of Athens. The Domination of the Acciaiuioli*] (Thessalonica, 2006), 189.

[48] Thiriet, *Assemblées*, I, 69, II 14, 100; Thiriet, *Régestes*, I 63, 94, 176, II 29, 105, 220; Mallett-Hale, *Venice*, 74–76.

[49] L. Thalloczy - C. Jerejek, *Acta et diplomata Res Albaniae Mediae Aetatis Illustrantia* (Vienna, 1913), 126; *Cronaca dei Tocco*, vv. 133, 315, 399, 633, 1509, 2497, 3666.

[50] Kantakouzenos, I, 174.

[51] Pseudo-Kodinos, 345, 348.

in early fourteenth-century Europe favoured the employment of small companies of mercenaries. The Italian cities in particular employed such soldiers extensively due to the availability of considerable liquid wealth.[52] However, some companies grew so strong that they demanded bribes from cities even during peace time.[53] While the 1370s and 1380s saw significant changes in the function of the companies in Europe mainly due to the growth of better organised state formations and changes in the economic conditions in Italy, after the 1370s there are no more references to Byzantine soldier companies. [54] This should be attributed to the extremely limited available information. It does not mean that the Byzantine state stopped employing small companies of soldiers.

In the second category of mercenaries belong troops hired for individual campaigns. Short-term contracts were very common also in Western Europe, particularly in Italy, where the city-states found it safer to hire mercenary companies through temporary contracts.[55] The employment of mercenaries on a casual basis was financially profitable, since the state had to pay them only for the services they provided during a particular military operation. This type of recruitment seems to have rather common in late Byzantium. In 1261, Alexios Strategopoulos, heading against Epiros, had been ordered to recruit locally in Macedonia and Thrace according to his needs.[56] In 1304, the general Sgouros was sent to Asia Minor with a few soldiers and a small amount of money to recruit locally.[57] Similarly, it seems difficult to believe that if the Catalans had succeeded in removing the Turks from Asia Minor, the Byzantine state would have been willing to continue keeping them in its service, given the huge cost of their service. In another instance, when in 1332 the Byzantine army in Bulgaria was cut off from Byzantium, on the eve of the battle of Rosokastro, Andronikos III told his troops 'we do not have allied cities that can

---

[52] Mallett, *Mercenaries and their Masters,* 25; idem, "Condottieri and Captains in Renaissance Italy," in D.J.B. Trim (ed.), *The Chivalric Ethos and the Development of Military Professionalism* (Leiden, 2003), 69.

[53] See W. Caffero, *Mercenary Companies and the Decline of Siena* (Baltimore, 1998), passim.

[54] Mallett, *Mercenaries and their Masters,* 51; Caferro, *Siena,* 13; Bartusis, *The Late Byzantine Army,* 205.

[55] Malett, "Condottieri," 68.

[56] Gregoras, I, 83.

[57] Pachymeres, IV, 455.

aid us today in this unexpected war, nor can we summon a mercenary army from somewhere.'[58] In 1349, the Byzantine governor of the Morea and son of John VI, Manuel Kantakouzenos recruited mercenaries from the region south of Vonitsa to crush the rebels in the Morea.[59] In 1359, in his attempt to claim the throne of the despotate of Epiros, Nikephoros Angelos Orsini hired and used against the Albanians in the battle of Acheloos a group of Turkish soldiers who at that moment were plundering Thessaly.[60] Other mercenaries hired for temporary service were those the Byzantine state borrowed from the Serbians. In 1313, Serbia sent 2,000 Cuman mercenaries to Byzantium to fight against the troops of the Turcoman adventurer Halil, who had previously allied himself with the Catalan Grand Company. According to Kantakouzenos, in 1320 Cuman mercenaries had not been returned to their paymaster, the king of Serbia and were employed by Byzantium.[61] In 1343, Kantakouzenos' army included German heavy cavalry mercenaries, provided and paid by the Serbian ruler.[62]

There are groups of mercenaries for which it is not certain whether they were hired for the purposes of a specific campaign or they performed long-term service. Nor do we know whether their remuneration consisted exclusively of cash payments or they received *pronoiai* or other grants of land. It is not known whether the Turkish mercenaries sent to the Peloponnese in the early 1260s were permanent troops or they were recruited only for the needs of the specific military operations. Pachymeres relates that Michael VIII sent in the Peloponnese 'the entire Persian force' implying that they were committed to long-term service.[63] According to the *Chronicle of the Morea*, Michael VIII hired Turkish soldiers only for this specific campaign. They had agreed to fight for one year and before the termination of their contract, they deserted the Byzantines complaining that they had not been paid for six months.[64]

---

[58] Gregoras, I, 485.

[59] Kantakouzenos, III, 88. For Manuel Kantakouzenos see Nicol, *The Byzantine Family of Kantakouzenos*, 122–129.

[60] Kantakouzenos, III, 319. Nikephoros Angelos Orsini was the son of the Italian count of Cephalonia Giovanni Orsini and of Maria, the daughter of Nikephoros of Epiros: Polemis, *The Doukai*, 99–100.

[61] Kantakouzenos, I, 35.

[62] Kantakouzenos, II, 356, 361.

[63] Pachymeres, I, 273.

[64] *Chronicle of the Morea*, vv. 4549–4555, 5099, 5722–5729; *Libros de Los Fechos*, ch. 359–362; *Livre de la Conqueste*, ch. 345–359.

We know nothing about the German and other Latin mercenaries hired by Byzantium in the fourteenth century apart from those Kantakouzenos borrowed from Serbia.[65] Despite the lack of sufficient evidence, it seems reasonable to conclude that the German mercenaries, who appear in the Byzantine sources, had come to Byzantium from Italy and Serbia. It has been argued that between 1320 and 1360 roughly 700 German cavalry leaders and 10,000 men-at-arms were active in Italy.[66] The increasing wealth of the Serbian rulers allowed them to hire German troops. It is logical that some of them sought further employment in Byzantium and it is probable that many European professional soldiers were attracted to the civil wars and the endemic warfare in the Balkans. It is impossible to determine whether the Germans were hired by the Serbians and the Byzantines were companies of soldiers. Nevertheless, this possibility cannot be excluded, particularly since most of the Germans fighting in Italy were members of companies of soldiers.

Furthermore, it is not clear whether the mercenary soldiers, who made up the units of the palace guard, were hired for permanent service. In addition, neither their specific function nor their importance during the late Byzantine period can be established with certainty. Pseudo-Kodinos states six distinct groups of palace guards: the Varangians, the Paramonai, the Mourtatoi, the Tzakones, the Vardariotai, and the Kortinarioi. The Varangians appear repeatedly in the sources, identified as axe-bearers.[67] According to the *Chronicle of the Morea,* the Varangians were guarding William II of Villehardouin during his captivity in Constantinople after his defeat at Pelagonia.[68] Kantakouzenos indicates that the Varangians accompanied the emperors on their campaigns, stating that it was customary for the Varangians to hold the keys of the cities wherever the emperor stayed. He adds that immediately after the death of Andronikos III he placed the deceased emperor's sons John and Michael under the protection of the Varangians.[69] Much less is known about the other palace guard units. Pseudo-Kodinos states that there were Paramonai with

---

[65] See Kantakouzenos, I, 141, 173, II, 166, 354.

[66] Mallett, *Mercenaries and their Masters*, 28–29; G.M. Varanini "Mercenari tedeschi in Italia nel Trecento: problema e line di ricerca," in S. De Rachewiltz – J. Riedman (eds.), *Kommunikation und Mobilität im Mittelalter* (Vienna, 1995), 159–178.

[67] Pachymeres, I,101,145,185; Gregoras, I, 303, 398, 566; Kantakouzenos, I, 200.

[68] *Chronicle of the Morea*, vv. 4319–4320.

[69] Kantakouzenos, I, 389, 560.

horses and Paramonai without horses, all bearing a sword.[70] Pseudo-Kodinos relates that the Mourtatoi were foot archers, while a western source states that they were of mixed Turkish and Byzantine origins.[71] The *Tzakones* were armed with clubs and wore distinctive breastplate with white lions, while the Vardariotai performed police duties.[72] The Vardariotai are attested in the sources during the reign of Manuel I Komnenos. Pseudo-Kodinos states that they were Turks established by an emperor in the Vardar area.[73] He refers to a resettlement of Hungarians (whom the tenth-eleventh Byzantine sources call 'Turks') in Macedonia along the Vardar valley during the tenth century.[74] According to Pseudo-Kodinos, who does not say anything about their weapons, the Kortinarioi were serving the imperial tent which was called *korte*.[75] Although these groups of soldiers composed the imperial guard, Pachymeres' account includes an exception. He states that Vardariotai and Kortinarioi served also Michael VIII's brother, the despot John.[76] This is an indication that John Palaiologos held a position higher than that of a despot and was placed just below the emperor and shows his influential role in the state affairs.

An important question is whether the palace guard units took part in the actual fighting. The available information does not exclude the possibility that occasionally palace guard units were active fighting units. In 1256, the Vardareiotai followed Theodore II on his campaign in Europe. Akropolites says that they had their own tents and they were under the command of their *primmikerios*.[77] It is certain that the Kortinarioi and Vardariotai who were in John Palaiologos' service accompanied him on his numerous campaigns. In 1329, the

[70] Pseudo-Kodinos, 180, 226. According to a patriarchical document dated in 1315 the Paramonai appear to perform police duties: *Das register des Patriarchats*, I, 170.

[71] Pseudo-Kodinos, 180; *Directorium ad faciendum passagium transmarinum*, C.R. Beazley (ed.), *AHR* 13 (1907–1908), 102; Weiss (*Kantakuzenos*, 69) discusses the possibility that the *Mourtatoi* were Christianised Turks.

[72] Pseudo-Kodinos, I, 181–182.

[73] Pseudo-Kodinos, 182; A. Hohlweg, *Beiträge zur Verwaltungsgeschichte des Oströmischen Reiches unter den Komnenen* (Munich, 1965), 62.

[74] Bartusis, *The Late Byzantine Army*, 280.

[75] Pseudo-Kodinos, 181. The Kortinarioi are attested in the tenth century treatises on imperial expedition compiled by Constantine VII as attendants of the *korti* which was the emperor's pavilion or the headquarters during a campaign: Constantine VII Porphyrogenitus, *Three Treatises*, 128, 165.

[76] Pachymeres, II, 417.

[77] Akropolites, I, 131; Pseudo-Kodinos (182) confirms that the head of the Vardareiotai was a *primmikerios*.

Varangians accompanied Andronikos III in his campaign on Chios. It is reasonable that in these cases, since the palace guards participated in the campaign, they could have been expected to fight. Otherwise, their presence would impose an additional burden to the logistical needs of the campaigning army. An argument in support of the view that the Vardareiotai campaigning with Theodore II fought on the battlefield is that Theodore II significantly increased the size of the army enlisting even those who had never been enrolled in the ranks of the army and those who served the emperors in the game preserves and in hunting game.[78] Since people who were not enrolled in the army were enlisted, logically the Vardareiotai, an armed palace guard unit, were also enlisted as a fighting force.[79] Moreover, in his description of the siege of Constantinople by Murad II in 1422, Doukas reports the presence in the imperial gate of Cretan guards who were faithful to the emperor.[80] These Cretans took part in the defence of the city against the Ottomans. It is impossible to assess the strength of the palace guard units, because we know nothing about their size. There is only a single reference to their numerical strength. Kantakouzenos says that, immediately after the death of Andronikos III in 1341 he installed in the palace a guard of 500 soldiers composed of his *oikeioi* and the *vasilikoi* (palace guards). Later he reduced the number of the palace guard to its customary strength.[81] This makes clear that in 1341 the palace garrison was made up of substantially fewer than 500 soldiers.

There is substantial information suggesting that large groups of mercenaries and troops composed of soldiers sharing a common cultural background were fighting under their own commanders. In 1255, while on campaign against the Bulgarians, Theodore II held a military council in which participated not only the Byzantine generals but also the commanders of the Latin and the Cuman soldiers.[82] In 1259, Nikephoros Rimpsas, who was of Turkish origin, though a Christian,

[78] Akropolites, I, 125.

[79] Hohlweg (*Beiträge*, 62) argues that the late Byzantine Vardariotai were not a fighting force.

[80] Doukas, 229.

[81] Kantakouzenos, II, 14.

[82] Akropolites (I, 120) states that the heads of the foreign mercenaries participated in the council due to the difficult situation caused by the severe weather conditions, which forced the army to halt its march. Therefore, Akropolites implies that it was uncustomary for the heads of the foreign mercenaries to participate in war councils.

participated in the battle of Pelagonia.[83] Apparently, this is the same Rimpsas who, according to Pachymeres, in 1273 led a contingent of Turkish troops in a campaign on Thessaly. Gregoras' account of the same campaign, does not mention the presence of foreign Turks but of the *Tourkopouloi*.[84] Since Rimpsas was a Christian and the *Tourkopoloi* were Christianised Turks, a by-product of Byzantine-Turkish contacts, it is quite possible that Rimpsas was the head of the *Tourkopoloi* in the battle of Neopatras.[85] In other examples, Chortatzes was the leader of the Cretan mercenaries established in Asia Minor at the end of the thirteenth century and Melik was the commander of the Turkish mercenaries employed by the Byzantines in the Morea in the 1260s.[86]

An argument in support of the view that the foreign mercenaries were commanded by their own leaders is that their fought with their own weapons. The Byzantine armies adapted their tactics according to the nature of the enemy, using the appropriate weaponry. For instance, they employed Turkish and Cuman light cavalry archers against the Latin heavy cavalry, and heavy cavalry against the Turks. Byzantine military commanders wanted to exploit the tactical advantages light cavalry archers would give them. These groups of soldiers fought under their own leaders and used distinctive battlefield tactics.

The sources say almost nothing about the process of employing individual mercenaries. Nonetheless, a reference letter Kydones wrote sometime between 1389 and 1390 to Manuel II on behalf of a mercenary from Piccardy might be characteristic of the career of a mercenary who did not belong to an organised company and of the process of employing soldiers in the late fourteenth century. Kydones reports that this soldier, following the custom of his fatherland spent a large part of his life far from his home helping those in need. He was hired by the Genoese ruler of Lesvos, Francesco Gattilusio and later by the semi-independent governor of Ainos, Nicolo Gattilusio, Francesco's brother. However, being dissatisfied by the treatment and low payment he received and being well-born and trained in warfare he wished to seek honour and fight among the best. For that reason he wished to fight for the Byzantine emperor.[87] It is impossible on the basis of a single surviving letter to assess the extent to which individual merce-

[83] Akropolites, I, 170.
[84] Pachymeres, II, 425; Gregoras, I, 111.
[85] Bartusis, *The Late Byzantine Army*, 61–62.
[86] *Chronicle of the Morea*, v. 5171; Pachymeres, III, 249.
[87] Kydones, *Correspondance*, II, 127; Tinnefeld, *Briefe*, IV, 143–145.

nary soldiers were hired in this manner. We do not know how often the emperor was involved in the recruitment of individual soldiers and it is likely that Kydones took advantage of his close friendship with Manuel II to introduce the aforementioned soldier. Nonetheless, that the emperor was involved in the process of enlisting mercenaries shows the simplification of the military administration of the later Byzantine empire. In the last decades of the fourteenth century the rulers of the small states in the Balkans could not afford the recruitment of large armies. Nor were they supported by an elaborate administrative infrastructure. Usually, they were personally in charge of recruiting mercenaries, who very often were bound not to the state they served but to the person of the ruler who employed them.

The officer who was supposed to be in charge of the Western European mercenaries was the *megas konostaulos*.[88] The creation of this office seems to have been a late Byzantine development and reflects the significant role of European mercenaries, which, however, had been rather important since the eleventh century, long before the period under discussion. It has been argued that the *megas konostaulos* was the successor of the *comes stabuli*.[89] It might be supposed that the title *konostaulos* was one that the Latin mercenaries would understand in terms of rank and responsibilities. One wonders whether it is not a coincidence that in the mid-thirteenth century the units of mercenaries serving in Italy started being called *conestabilia* or *conestableria* and their commander *conestabilis*, and that these terms came to Italy through France.[90] Moreover, archival evidence shows that small companies of mercenaries, who were manning Venetian fortresses in Frankish Greece, were under the command of *conestabili*.[91] In this case the office of *megas konostaulos* is not related to that of *komes tou staulou* (*comes stabuli*), which is attested for the last time in the sources in late tenth century and its holder was responsible for the imperial military stables.[92]

---

[88] Pseudo-Kodinos, 175; Sphrantzes, 128.

[89] R. Guilland, *Recherches sur les institutions byzantines*, 2 vols. (Amsterdam, 1967), I, 469.

[90] D. Waley, "The Army of the Florentine Republic from the Twelfth to the Fourteenth Century," in N. Rubinstein (ed.), *Florentine Studies. Politics and Society in Renaissance Florence* (London, 1968), 83–84.

[91] Thiriet, *Assemblées*, III, 100; *MP*, 215; Mallett-Hale, *Venice*, 76–78.

[92] N. Oikonomides, " L' évolution de l'organisation administrative de l' empire byzantin", *TM* 6 (1976), 145.

Moreover, the Byzantines long before 1204 knew this office as a foreign institution. Describing the military conflict between Alexios I Komnenos (1081–1118) and the Normans in the 1080s Anna Komnene states that one of the generals of the troops of Bohemond was the *megas konostaulos* Brienne.[93] However, she does not specify the functions of this office. Instead, she refers to Byzantine commanders of Frankish mercenary troops without calling them *megas konostaulos.*[94] Therefore, the future emperor Michael VIII appointed *megas konostaulos* by John III is the first known Byzantine holder of this office, but not the first Byzantine commander in charge of Latin mercenaries.[95] It is worth noting that according to Pachymeres, it was an old custom for the *megas konostaulos* to be in charge of the 'Italian forces and subjects.'[96] This statement implies that before Michael Palaiologos there were earlier holders of the office, who are unattested by the available evidence. The sources also do not provide information, which could indicate differences in the role of the *megas konostaulos* and the earlier heads of Latin mercenaries, like those mentioned by Anna Komnene, in the military administration of the empire. Therefore, it is impossible to assess to what extent the introduction of the office of *megas konostaulos* reflects an administrative innovation.

Byzantine attitudes towards mercenaries

Many Western European humanists and authors have depicted the mercenaries as degraded soldiers. Cruelty, fickleness and greed are the main characteristics attributed to them.[97] Did late Byzantine authors

---

[93] Komnene, Anna, *Alexias*, D.R. Reinsch – A. Kambylis (eds.), (Berlin, 2001), 157.

[94] *Alexias*, 126, 211, 252. The commanders are Constantine Oumbertopoulos and Adrianos Komnenos. It should be stressed that in these instances she is not referring to the Varangians.

[95] Akropolites, I, 134.

[96] Pachymeres, I, 37.

[97] For attitudes towards mercenaries in the West see J. France, "Mercenaries and Paid Men. The Mercenary Identity in the Middle Ages. Introduction," in J. France (ed.), *Mercenaries and Paid Men. The Mercenary Identity in the Middle Ages*, (Leiden, 2008), 1–13; D. Crouch, "William Marshall and the Mercenariat," in J. France (ed.), *Mercenaries and Paid Men. The Mercenary Identity in the Middle Ages*, (Leiden, 2008), 16–32; M. Mallett, *Mercenaries and their Masters*, 196–209; Caffero, *The Decline of Siena*, 2, 86–87; idem, "Slaying the Hydra-Headed Beast: Italy and the Companies of Adventure on the Fourteenth Century," in L.J.A. Villalon – D.J. Kagay (eds.), *Crusaders, Condottieri and the Cannon. Medieval Warfare in Societies around the Mediterranean* (Leiden, 2003), 287–288.

saw mercenary soldiers as disloyal, ineffective and rapacious bands of soldiers? The first Byzantine author to comment on the employment of mercenaries after 1204 was the emperor and prolific author Theodore II Laskaris. In a treatise he compiled in the early 1250s and before his accession on the throne (1254), in which he discusses the relationship between the ruler and his subjects, Theodore II thought that hiring mercenaries was an imperial duty aimed at the defence of the empire.[98] However, after his accession he began to express doubts as to the expediency of this policy. In a letter addressed to his old teacher Nikephoros Blemmydes, the emperor points out the inevitable high cost of the maintenance of military forces. He wonders whether he could reduce the size of the army or the amount of money required for the maintenance of military forces. He concludes that either action would benefit the enemies of the empire. Theodore remarks that neither the Turks, nor the Bulgarians, nor the Italians, nor the Serbians would ever help the Byzantines. Instead, the Byzantines should rely on their own resources. This statement shows that Theodore II is sceptical of the high cost of his foreign mercenaries.[99]

Theodore II's letter is related to the dispute over the imposition of new taxes or the increase of old ones, for the implementation of his reforms that focused on the increase of the size of the army.[100] The emperor argues that due to John III's military achievements, the empire stretched from Rhodes to Dyrrachion and therefore, it was essential to maintain a successful conquering army. Then he defended his policy of increasing the size of the army by recruiting native soldiers and objected Blemmydes' proposal that money hoarded in the treasury should be used instead of taxes for the implementation of this policy.[101] It seems that Theodore II's military policies were expensive. The increase in the size of the army, that Laskaris' army spent much time campaigning and that Anatolian armies were stationed in Europe for more than one year, must have resulted in the imposition of heavy taxation. Whatever reservations Theodore II had about their employment, Latin mercenaries participated in his campaigns and played an influential role in the events following his death. Michael Palaiologos relied on them in order to usurp the throne. Probably,

[98] Theodore II, *Opuscula*, 126.
[99] Theodore II, *Epistulae*, 58.
[100] For the dispute between Blemmydes and Laskaris over imperial taxation see Angelov, *Imperial Ideology*, 292–296.
[101] Theodore II, *Epistulae*, 57–58; Angelov, *Imperial Ideology*, 295.

even if Theodore II had a plan to reduce the number of foreign mercenaries, it did not materialise.

Pachymeres and Gregoras are critical of the large groups of mercenaries which were employed by Andronikos II in the 1290s and in the first decade of the fourteenth century.[102] Pachymeres is critical of the government's decision to impose a 10% levy on the *pronoiai* to fund the Cretan mercenaries Andronikos II recruited sometime before 1295 and settled in Asia Minor.[103] In 1295, the Cretans played a leading role in the events surrounding Alexios Philanthropenos' rebellion in Asia Minor. Pachymeres does not seem to be critical of the Cretans' attitude, who despite initially being staunch supporters of Philanthropenos, captured and handed him to the imperial authorities after being promised immunity and rewards. Instead, he remarks that Philanthropenos' indecisiveness and cautiousness made the Cretans feel unsafe and fear that he would change his mind and reach an agreement with the emperor.[104] Consequently, Pachymeres does not see the Cretans as a disloyal band of mercenaries. For him they deserted Philanthropenos not because they were greedy mercenaries but because they feared that if Philanthropenos made a deal with the emperor, they would have found themselves in a rather difficult position.[105]

As is the case with the Cretans, Pachymeres' analysis of the events following the employment of the Alans in 1301 reflects his opposition to the financial and military policies of Andronikos II. He writes that Andronikos II had prepared supplies for the Alans through public contributions and provided them with horses and weapons, which he took away from native soldiers. He disagrees with the employment of the Alans at the expense of native troops.[106] For Pachymeres that the emperor provided the Alans with resources that previously belonged to native Byzantine soldiers, contributed to the defeat of the Byzantines at the battle of Bapheus (27 July1302) at the hands of the Ottoman Turks. The Byzantine army that fought at this battle was composed of Byzantine and Alan soldiers under the command of Leo Mouzalon.

[102] For Byzantine reactions to the employment of foreign mercenaries during Andronikos II's reign see S. Kyriakidis, "The Employment of Large Groups of Mercenaries in Byzantiumin the Period ca. 1290–1305 as Viewed by the Sources," *B* 69 (2009), 208–230.

[103] Pachymeres, III, 235–237.

[104] Pachymeres, III, 247– 249.

[105] For Livadarios see *PLP*, 14859.

[106] Pachymeres, IV, 339.

Pachymeres remarks that although the native Byzantine soldiers were expected to fight wholeheartedly for the defence of their homes, they did not do so. They were demoralised and unwilling to fight because of the money and weapons that had been taken away from them to supply the Alans. Pachymeres' account of the battle of Bapheus does not portray the Alans as undisciplined mercenaries interested only in plunder. He comments that the Alans proved themselves useful; they encircled the enemy but they suffered heavy casualties.[107] Therefore, Pachymeres attributes the Byzantine defeat at Bapheus to mismanagement of resources.

In 1302, Andronikos II hired the Catalan Grand Company, which was formed during the wars in Sicily between the Aragonese and the Angevins and was partly made up of Almugavars. After the end of the war in Sicily, they sought employment in Byzantium. Pachymeres blames the government for mismanaging the Catalans, as he does with the Cretans and the Alans. He argues that Andronikos II hired the Catalan Grand Company out of desperation and without taking into serious consideration the state's finances, which were insufficient to maintain such an expensive group of soldiers. More specifically, Pachymeres claims that in order to produce the cash to afford the payment of the Catalans and to avert them from ravaging imperial territories, a special levy on grain was imposed, which was collected from the Balkan regions of the empire. Furthermore, the government confiscated one third of the *pronoiai* located in the western territories, discontinued the payment of the servants of the palace and devalued the currency.[108] Pachymeres is also critical of the extraordinary authority the Catalans were invested with. He relates that the leader of the Catalans, Roger de Flor became the supreme commander of the Byzantine army (*strategos autokrator*) who arranged the military affairs of the empire and the salaries of the soldiers according to his wishes.[109]

The account of Ramon Muntaner of the agreement between the Catalans and the emperor confirms Pachymeres' claims that the Catalans were offered too much. Muntaner, who states that he participated in the drawing of the agreement between the Company and

[107] Pachymeres, IV, 367.
[108] Pachymeres, IV, 539–541.
[109] Pachymeres, IV, 465.

Byzantium, relates that it was agreed that the emperor should give pay for four months to all those Roger de Flor could bring along and he should keep them in his pay as long as they wished to stay.[110] This implies the absence of any agreement defining the number of troops and length of service of the Catalans and contradicts Andronikos II's claim, as mentioned by Pachymeres, that according to the *chrysobulls* he had sent Roger de Flor, the Catalan force should be made up of 1,000 infantry and 500 cavalry and not of 6,000 in total.[111] Therefore, we cannot be sure about the exact provisions of the agreement between Andronikos II and the Catalan concerning the number of soldiers provided, the duration of service, pay clauses, division of booty, fate of prisoners, provision of supplies and jurisdiction in the case of lawbreaking. Nevertheless, Pachymeres' complaints and Muntaner's account indicate that even if there was an initial agreement over these issues it was rather favourable to the Catalans. This can also be concluded through a comparison of the accounts of Pachymeres and Muntaner with contemporary contracts agreed between Italian cities and companies of soldiers.[112] Although these contracts varied in length and content, certain matters are mentioned in almost all of them. The number of soldiers employed is far smaller than the size of the Catalan Grand Company and normally the length of service is quite specific and stretches from two months to one year. Furthermore, the usual arrangement was prisoners and booty to be given to the authorities of the contracting state. The captor was to receive compensation. The contracts also fixed the number of horses each mercenary could possess and some of them included details about armour and weaponry. Usually, there was a division of jurisdiction between the employing authority and the company's commanders. The company's commanders were in charge of the disputes within the company but they did not have jurisdiction over the local population.[113] Pachymeres does not say for how long the Catalans were hired by the emperor, while Muntaner writes that they could stay in Byzantium's service for as long as they wished to. Furthermore, more troops arrived after the initial agree-

---

[110] Muntaner, II, 423.

[111] Pachymeres, IV, 533.

[112] For the contracts agreed between the Italian states and companies of mercenaries see D. Waley, "Condotte and Condottieri in the Thirteenth Century," *Proceedings of the British Academy*, 61 (1975), 337–371.

[113] Waley, *Condotte* and *Condottieri*, 340–342.

ment with the Company and the Catalans seemed to have jurisdiction over local governors. With regard to the booty it is likely that there was an initial agreement for which we do not have any details. Muntaner writes that after their first victory over the Turks the Catalans sent most of the booty to the emperor.[114]

Pachymeres does not express any doubt about the military value and competence of the Catalan company. He portrays Roger de Flor as someone who had proven his bravery and 'was minded towards military matters.'[115] Pachymeres also comments that the rumour of the forthcoming Catalan attack was sufficient to make the Turks of Karasi lift the siege of Germe and flee without order.[116] Nevertheless, he attacks the attitude of the Catalans. He blames them for being unwilling to fight and for staying inactive in Kyzikos, where they arrived in October 1303, with the excuse that they had not received their salaries.[117] According to Pachymeres, their presence burdened the local population against whom the Catalans committed hideous crimes. He argues that although by the end of March 1304 they had received their salaries, the Catalans did not campaign. Instead, they were involved in fighting with the Alans.[118] Being critical of the way the government handled its mercenaries and seeing financial profit as the main motivation of soldiers, Pachymeres points out that one of the causes of the conflict between the Catalan and the Alan mercenaries was the reaction of the latter to the much higher financial rewards the Catalans received.[119] Muntaner, as a member of the Catalan Company sees things differently. He says that although the Company was ready to depart from Kyzikos on 1 November 1303 the severe weather conditions prohibited them from campaigning until April 1304. Thus, the Catalans were forced to remain in Kyzikos. Nonetheless, Muntaner admits that their maintenance imposed a heavy burden on the local population.[120]

Furthermore, using economic reasoning to explain the attitude of soldiers, Pachymeres portrays greed and rapacity as the main incentives of the Catalans. For instance, he claims that at some point the

[114] Muntaner, II, 430–431.
[115] Pachymeres, IV, 431; Gregoras, I, 218.
[116] Pachymeres, IV, 467.
[117] Failler, "Chronologie et composition," 55.
[118] Pachymeres, IV, 433, 459–461.
[119] Pachymeres, IV, 463.
[120] Muntaner, II, 432–433.

Catalans requested to be paid not only their regular salaries but also for their 'supposed previous activities.' He sees this as an unfair demand because, as he comments, the Catalans asked to be paid for a period during which they did not fight. Pachymeres adds that when they received their payment the Catalans left Mytilene and instead of attacking the Turks they plundered Madytos.[121] Similarly, he relates that Berenguer d' Entença arrived in the empire because he was promised by Roger de Flor that he would receive great rewards from the emperor.[122] Moreover, Pachymeres blames the Catalans for blackmailing the emperor. As he writes, the Catalan envoys told Andronikos II that if he would pay the salaries they demanded, they would refrain from ravaging the empire and they would be ready to fight whenever the emperor asked them.[123] These statements vividly contrast with Muntaner, who writes that Berenguer d' Entença had sworn brotherhood with Roger de Flor and came to Byzantium leading his company of 300 horsemen for the emperor's honour and for the love of Roger de Flor. Moreover, Muntaner justifies the Catalan attacks as the result of the bad faith of the emperor, who was planning to have them killed after restoring Byzantine control over Asia Minor and as vengeance for the murders of Roger de Flor and Catalan envoys.[124]

To emphasise his disapproval of the employment of the Catalans, as well as their actions Pachymeres quotes Plato. He writes

> Plato correctly outlines the mercenary service in the *Laws*. He said that most of them become arrogant, unjust and violent and the most imprudent people apart from very few. Of all the Cardinal Virtues, which are possible to be witnessed in the citizens involved in civil wars, only the fourth one, which someone could call bravery, can be attributed as perfect to mercenaries because they wish to die in war only, not because of virtue, but only for the sake of salary; and for this, they are easily involved many times in reckless violence, being arrogant and against justice because they are unjust, arrogant against prudence because they are violent and being totally imprudent because of the absence of prudence.[125]

[121] Pachymeres, IV, 529.
[122] Pachymeres, IV, 531.
[123] Pachymeres, IV, 549.
[124] Muntaner, II, 442–443, 450–451.
[125] Pachymeres, IV, 529: (cf. Plato, *Laws*, I, 630 b).

Furthermore, Pachymeres' views about the Catalan Grand Company might have been affected by the significant differences between the Catalans and the other mercenaries employed by Andronikos II. The most obvious difference was the intense conflict between Byzantium and the Catalan Grand Company and the devastation the Catalans inflicted on the Byzantine state. Therefore, it was inevitable that they would be subjected to harsh criticism. Moreover, they were a self-interested company of mercenaries. Their aim was to seek maximum financial profit through war. They were not refugees like the Alans and the Cretans, who in return for land to settle provided the army with mercenary soldiers. Nor can the Catalan Grand Company be compared with the small groups of soldiers or individual mercenaries employed by Byzantium. It was difficult for a state like early fourteenth-century Byzantium, which experienced severe financial difficulties, internal conflicts and territorial reduction, to control a large mercenary company with its own internal organisational system and hierarchy.

Gregoras provides a shorter account of the foreign mercenaries employed by Andronikos II than Pachymeres does. He depicts a more positive image of Andronikos II, whom he supported in the civil war of 1321–1328. Moreover, Gregoras' approach to history differs from that of Pachymeres. Influenced by classical models he considers that although humans make conscious choices, often it is Fortune (*Tyche*) that determines outcomes.[126] Consequently, Gregoras does not adopt Pachymeres' economic analysis to explain the political and military developments of the period. Furthermore, he was writing in the mid-fourteenth century from a longer chronological distance than Pachymeres.

Avoiding criticising Andronikos II's policies, Gregoras is critical of the Cretan mercenaries. He states that although they received great honours from Philanthropenos, 'they did not wish to remain in the established situation, but [the Cretans] had thoughts of things that were much higher than their value and fortune.' He explains the attitude of the Cretans stating that long and continuous success leads mindless people to insolence.[127] Nevertheless, Gregoras like Pachymeres observes that Philanthropenos' hesitations made the

[126] Angelov, *Imperial Ideology*, 256.
[127] Gregoras, I, 196.

Cretans feel unsafe and that they warned him that if he delayed to proclaim himself emperor, he would be betrayed.[128] However, unlike Pachymeres, he is rather critical of the fact that the Cretan mercenaries captured Philanthropenos after receiving bribes and promises for great rewards. For Gregoras the Cretans were mainly motivated by greed and ambition and not by the threat of being isolated and abandoned by Philanthropenos.[129] Therefore, by blaming the Cretans for the rebellion of Alexios Philanthropenos Gregoras avoids criticising Andronikos II.

Gregoras says little about the Alan mercenaries. He pays particular attention to the high cost of their employment and to the raids they carried out on Byzantine soil. This contrasts with Pachymeres who does not omit to state their depredations but emphasises the state's mishandling of the Alans. For Gregoras the Alans played a central role in Michael IX's failure in Magnesia. He presents them as undisciplined soldiers whose attitude demoralised the Byzantine camp. He also comments that by plundering Byzantine possessions the Alans facilitated the Turkish advance. As he writes, once the Turks attacked, the Alans did not fight but fled.[130] Gregoras says nothing about the provision of the Alans with resources that had been taken away from the native soldiers. Consequently, he does not follow Pachymeres in criticising the way the government handled the Alans. Instead, he presents them as being incompetent and undisciplined soldiers. Thus, Gregoras does not blame the government for mishandling the Alans.

Gregoras' account of the Catalan Grand Company differs from Pachymeres' analysis. Writing in the mid-fourteenth century, when the employment of companies of mercenaries in Europe reached its peak, and seeing things from a longer chronological distance than Pachymeres does, Gregoras is well aware of the unsafe situation in which mercenary companies find themselves during peace. As he writes,

> When the kings made peace, to those [Gregoras implies the Catalan Company] of Frederick (Frederick III of Aragon) occurred the need to consider where to go and find profit. They did not have houses and properties to speed them to return. But each of them being from a different place and many from many places and being without furniture and

[128] Gregoras, I, 199.
[129] Gregoras, I, 201.
[130] Gregoras, I, 205–206.

> clothing, for the sake of profit from plunder they were working together and they were nourished through a life of wandering across the seas.[131]

The devastation they caused does not prevent Gregoras from pointing out the military competence and professionalism of the Catalans. More specifically, Gregoras relates that the Turks lifted the siege of Philadelphia because,

> The enemy seeing the well-ordered march of the Latins, the splendour of their weapons and their unmoved eagerness, they fled. And they carried themselves away, not only far away from the city but close to the old Roman frontiers. This was the quantity and the quality of that army [the Catalan Grand Company], and it was so exceptionally organised because of the arms, the experience acquired at wars and because of its multitude (alongside the Latins campaigned not only the select of the Romans but also as many as were available from the army of the Alans). And because of these, they [The Catalans] inflicted such terror on the enemy that many said then, that if the emperor's orders motivated by cowardice did not prevent them from marching further, there would be nothing to prevent all the Roman cities and lands, which were clearly in the hands of the enemy, from being handed to the emperor within a short period of time.

Gregoras continues by stating that since they were not provided with guides, the Catalans decided not to advance further and while the Byzantines and the Alans returned to their homes the Catalans having nowhere to go became a burden to the Byzantine cities. Claiming that they had not received their payment, they plundered Byzantine possessions. Without going into detail he follows Pachymeres' criticism of the financial implications of the employment of the Catalans pointing out that the expenses for their maintenance emptied the imperial treasury. Gregoras attributes the failure of the employment of the Catalans to Divine intervention. He comments that those who believed that the Catalans would have succeeded in recovering Byzantine land, if they had not been stopped by imperial orders said so because they were,

> People who could see only the current situation and were not able to conceive anything beyond it. For in reality, this was a Divine decision taken long ago; that the Roman affairs will reach the extreme of misfortunes.[132]

---

[131] Gregoras, I, 219–220.
[132] Gregoras, I, 220–222.

This statement reflects Gregoras' views about Fortune. It was Fortune and Divine intervention that foiled the plans for the defence and recovery of Asia Minor and not the mistakes of the government, which Gregoras presents as things 'said by many' and not by him. He did not oppose Andronikos II's decision to employ the Catalans. Moreover, the pessimistic view that the disastrous effects of the employment of the Catalans were a Divine decision taken long ago reflects that Gregoras was writing in a period when Byzantium had not recovered from the civil wars of the 1340s and 1350s.

In his treatise, *On Kingship* (*Peri Basileias*) which he compiled around 1304 Thomas Magistros argues that the state should maintain sea and land military forces made up of native soldiers. He is negative towards foreign mercenaries, commenting that since ancient times the Byzantines used to employ foreign troops believing they would strengthen their armies. However, they prove to be weaker than is generally believed. He adds that the foreign mercenaries are not loyal soldiers and although they swear allegiance to the Byzantines, in reality they remain loyal only as long as the Byzantines are victorious against their enemies. If, due to fortune, the enemy proves stronger than the Byzantines, the foreign mercenaries desert Byzantium and join the enemy making them stronger and more courageous against the Byzantines filling the Byzantine army with fear and cowardice. Instead, the army should rely on native troops who should be provided with *pronoiai*, which should be hereditary at least for one generation for families of soldiers killed on the battlefield.[133]

Magistros does not believe that the employment of mercenaries satisfies the need of the Byzantine army to be constantly ready for war. Instead, he argues that the army should be composed of well-trained native soldiers and that all citizens should be trained in arms. [134] Magistros is the only known late Byzantine author who makes such a suggestion. Unlike Western European states, the late Byzantine empire did not establish urban militia or any other form of universal or territorial obligation to military service. Magistros might have been influenced by the specific circumstances of the conflict between Byzantium and the Catalans. The defence of the Byzantines against the Catalans in Macedonia relied on the strength of their fortifications and not on

[133] *PG* 145, col. 461.
[134] *PG* 145, col. 509.

successes of the imperial army on the battlefield. It is also possible that the failure of the Byzantine army to confront the Catalans led him to view the civilian population as a resource of manpower.[135] In addition, the failure of the employment of the Catalan Company and the Alans to bring any fruitful results reinforced Magistros' opposition to the employment of foreign soldiers.

Magistros' opposition to the recruitment of foreign mercenaries is also in agreement with his criticism of imperial taxation. The employment of large groups of mercenaries resulted in the imposition of heavy taxation and in his writings Magistros expressed strong ideas against tax increases reaching the point of objecting the need of tax collection.[136] Moreover, Magistros comments that the army should be composed of property owners who are well established in the city, where, as he writes, the tombs of their ancestors are. Those who own nothing, he concludes, have nothing to protect and they would easily submit in the manner of traitors for the sake of profit.[137] Consequently, another reason why Magistros opposes the employment of foreign mercenaries is that they had no links to the local society, although it is not unlikely that many of them invested their salaries on buying land in Byzantium. That Magistros' *On Kingship* borrowed phrases and ideas from Synesius' of Cyrene *On Kingship* (ca. 370–ca. 413) might have influenced his views about foreign troops. Synesius had suggested to the emperor Arcadius (395–408) to rid the court of Constantinople of the influence of German generals. Therefore, it is noticeable that both Synesius in the fifth century and Magistros in the fourteenth opposed the employment of foreigners.[138] In conclusion, Magistros disapproves foreign mercenaries mainly because they do not fit to his proposals about the fiscal and military organisation of the state.

Magistros' views about mercenaries can be contrasted to the ideas expressed by Theodore Metochites who in his writings approves the employment of foreign soldiers. Unlike Magistros, who opposed imperial taxation, Metochites argues that it is impossible to rule the empire without money and that it necessary the emperor to spend liberally money for the maintenance of indigenous and foreign soldiers for the defence of the empire against its greedy enemies. Metochites concludes

---

[135] Bartusis, *The Late Byzantine Army*, 308.
[136] See Angelov, *Imperial Ideology*, 298–303.
[137] *PG* 145, cols. 521–524.
[138] Angelov, *Imperial Ideology*, 190.

that it is necessary the throne to possess ample financial resources to fund a large army even during peace.[139]

Leading the Byzantine army in Macedonia in order to recover the cities captured by the Serbians during the civil war of 1341–1347 Kantakouzenos came face to face with the Serbian emperor Stefan Dušan. In this meeting Kantakouzenos complained that it was injustice the Serbian ruler to demand Byzantine cities as reward for the support he had given him during the civil war of 1341–1347. He stated that to the Byzantines it was a great dishonour to serve for pay instead for the love of honour and friendship.[140] Earlier, describing the entrance of Andronikos III in Constantinople and the deposition of Andronikos II in 1328, Kantakouzenos claims that he replaced the German soldiers who were to climb first the ladders to enter Constantinople with Byzantines, so that the glory would belong to the Byzantines.[141]

However, describing Andronikos III's preparations for the civil war against Andronikos II in 1321, Kantakouzenos comments that, 'the soldiers were eager to fight, and particularly those of the Latins who were Germans, and there were some of noble birth among them.'[142] In 1341, foreign mercenaries played a leading role in the coronation of Kantakouzenos in Didymoteichon. As he writes, he received one of the imperial boots from closest relatives and the other one from the most noble and illustrious of the Latin mercenaries, to some of whom, following the customary things, he awarded the title of *kavallarioi*.[143] It would be very interesting if Kantakouzenos was willing to provide more details about this ceremony given the importance and variations in the ceremony of dubbing to knighthood in Western Europe.[144] Moreover, as has been seen above, throughout his account Kantakouzenos points out that the constant readiness and mobility of mercenaries made them an essential part of the Byzantine army and his account of the construction of a siege tower under the instructions of a German mercenary shows that he appreciated the advanced technological knowledge possessed by these foreign soldiers.

---

139 Metochites, *Miscellanea*, 538–544.
140 Kantakouzenos, III, 143.
141 Kantakouzenos, I, 301.
142 Kantakouzenos, I, 98.
143 Kantakouzenos, II, 166.
144 See Keen, *Chivalry*, 64–82.

Kantakouzenos does not doubt the loyalty of mercenaries. He relates that when in 1343 the civil war started turning to his favour, his ally, Stefan Dušan began to fear that he would lose control over Kantakouzenos' movements, since he would not need Serbian military support any longer. Stefan Dušan, therefore, recalled the German heavy cavalry mercenaries he had given Kantakouzenos. The Germans accused the Serbian ruler of being malign and assured Kantakouzenos that, in case of a Serbian attack, they would fight on his side. They stated that this was the proper thing to do and that they could not be blamed for doing so. They justified that decision saying that if they were ordered by their paymaster (in this case the Serbian ruler) to aid someone, they would obey their master's order to desert him only if he were in a safe position. However, since they received the order to desert Kantakouzenos, while he was on campaign and ready to fight a battle, to desert him would be treachery, which they considered a shameful act. Shortly afterwards, when Kantakouzenos reached Veroia he discharged his German mercenaries and sent them to the Serbian ruler after rewarding them for their favour.[145] Kantakouzenos writes that when in 1350 his forces captured Veroia from the Serbians he fought against the Germans who had supported him seven years before. He claims that due to their previous loyalty and support he donated them money and horses and sent them to the Serbian ruler.[146] It is hard to believe that Kantakouzenos came across the same Germans who had borrowed from the Serbians seven years ago. Most probably with this statement Kantakouzenos wishes to promote his military profile as a gracious leader. Nevertheless, this is another indication of the prominent role played by Western mercenaries in the conflicts of the period.

In 1352, Kantakouzenos recruited 300 Catalan soldiers who belonged to the Catalan fleet which arrived in Bosporus to reinforce the Venetians in the conflict against the Geneose. Kantakouzenos concludes that the Catalans displayed great bravery on the battlefield.[147] They became a sort of bodyguard of Kantakouzenos. They were bound to his person and not to the imperial authority. When in 1354 John V entered Constantinople and forced Kantakouzenos to abdicate, the Catalans

---

145 Kantakouzenos, II, 356, 361.
146 Kantakouzenos, III, 125.
147 Kantakouzenos, III, 228.

occupied the fortress of the Golden Gate and declared their loyalty to Kantakouzenos and their determination to fight for him. Praising their loyalty and wishing to promote the idea that he abdicated voluntarily, Kantakouzenos writes that he found it hard to persuade them to surrender the fortress to John V Palaiologos.[148]

Therefore, Kantakouzenos expresses no doubts about the loyalty of foreign mercenaries and regards them an essential part of the Byzantine army. Then why did he call the service for pay slavish and in 1328 prevented the German soldiers from being the first to enter Constantinople? This contradiction is explained by the position of Kantakouzenos in Byzantine society. As a member of the military aristocracy he belonged to a social group which promoted the idea that fighting for an aristocrat was above all, a social obligation and duty; not a profession. This group had also developed a set of ethical ideals such as fighting for *patria*, faith, friendship, honour. A Byzantine emperor and aristocrat who promoted these ideals could not express positive views about soldiers who lacked any political and social ties with their employers and their only incentive to fight was financial profit.

However, Kantakouzenos implies that his mercenaries did not lack ties with the Byzantine society and that they displayed noble behaviour. The reasons why the German mercenaries did not abandon him when they were recalled by their paymaster are indicative of this. The prominent role of European mercenaries in his coronation shows that Kantakouzenos viewed them as members of the nobility. Similarly, by conferring upon them the title of *kavallarioi*, Kantakouzenos regarded these mercenaries as noble who shared his ideals and had political and personal ties with him and the empire. In addition, and perhaps most importantly, Kantakouzenos was a general with significant battlefield experience and knew that the logistical needs of war and battlefield reality made the employment of professional soldiers, who would fight for money, essential. Kantakouzenos' attitude towards mercenary service reflects another reality; that plunder and booty were no less attractive to mercenaries than to high aristocrats as incentive to risk their lives. This development was not restricted to Byzantium. In

[148] Kantakouzenos III, 286–287, 290–293, 300–304; Gregoras (III,151) comments that, these Catalans were light-armed beggars who were hired by Kantakouzenos because he mistrusted the Byzantines. Gregoras made these comments when he started expressing negative views about Kantakouzenos in his account.

Western Europe there were mercenary leaders who were knights and it has been concluded that in terms of motivation and conduct the line between noblemen and mercenaries it was difficult to draw.[149] This shows that for Kantakouzenos aristocratic attitudes towards warfare did not disapprove mercenary service and that he did not see mercenaries as ill-disciplined, rapacious and disloyal groups of soldiers.

In his addresses to the despot of the Morea, Theodore II Palaiologos and to the emperor Manuel II, George Gemistos Plethon proposed measures aiming at the re-organisation of the Byzantine Peloponnese which was threatened by the increasing strength of the Ottoman Turks. Plethon argues that the population of the Morea should be divided into soldiers and taxpayers because the two functions cannot be combined. As a result of this reform the army will be psychologically sound and will have high morale. The despot should get the army rid of mercenaries because they fulfill neither qualification.[150] In his address to Manuel II, Plethon is critical of the fact that although the taxes were small, they were numerous and collected in money and not in kind. Gemistos also rejected the raising of taxes to hire foreign mercenaries for the defence of the Peloponnese. He writes,

> It is absurd to tax each household to pay for mercenaries to defend the Isthmus. Taxation is already ruinous without this extra burden. In no case are mercenaries sufficient to defend Isthmus without reinforcement by our own troops who are unarmed and untrained.[151]

Although he appears critical of the employment of mercenaries, by promoting the idea of dividing the population into soldiers, who would not pay taxes and tax-payers Gemistos promoted the idea of creating professional full-time soldiers. Moreover, his opposition to the employment of foreign mercenaries is connected to the influence of classical authorities and to the patriotic spirit which features predominantly in his works. Ignoring the Albanians who immigrated to the Morea and reinforced the Byzantine army in the fifteenth century, the Slavic tribes which were living in the Peloponnese and that the armies of the various principalities established in Frankish Greece were predominantly mercenary in character, Gemistos claims to Manuel II

[149] Caferro, *Siena*, 5–6; Keen, *Chivalry*, 231.

[150] *PP*, IV, 130–132; Woodhouse, *George Plethon Gemistos*, 97.

[151] *PP*, III, 252–253; Woodhouse, *George Plethon Gemistos*, 103, whose translation this is.

that the Peloponnese had been 'the home of the Hellenes, and of no one else, since time immemorial.'[152]

The views of the marquis of Montferrat, Theodore Palaiologos are affected by the activities of the companies of soldiers fighting in fourteenth-century Italy and are similar to those expressed by contemporary Italian authors.[153] The marquis of Montferrat states that the mercenaries are cruel people with evil conscience interested only in profit. He exhorts the ruler not to employ mercenaries who speak one language exclusively. Furthermore, he suggests that the ruler not give the mercenaries important offices in recently conquered areas, since they will appropriate and keep these lands for themselves, betraying their paymaster.[154] This observation gives the impression that Theodore does not only describe the realities of contemporary Italian warfare, but also bases his argument on the conflict between Byzantium and the Catalans, having in mind the large size of the Catalan company and the Alan forces, and that members of the Catalan Grand Company had received titles and privileges. Nevertheless, Theodore admits that, since it is not possible for a ruler to defend his lands relying exclusively on his own men, the employment of mercenaries is necessary and suggests the employment of foreign troops in frontier regions, stating that they do not have any reservations in destroying the fields of the enemy.[155] It is also worth noting that Theodore's successor, the marquis John II Palaiologos was one of the first employers of the company of John Hawkwood, which was one of the most famous bands of professional soldiers that fought in fourteenth-century Italy. In 1361, the marquis himself led Hawkwood's company over the Alps.[156]

Consequently, to understand attitudes towards foreign soldiers it is necessary to take into account the wider context of the sources. For instance, Pachymeres is critical of Andronikos II and throughout his account, he explains past events and attitudes of soldiers using economic reason. For him, rapacity is the exclusive characteristic of the Catalans. In the same manner, the main cause of the failure of the foreign mercenaries to bring military success was the mismanagement

[152] Woodhouse, *George Plethon Gemistos*, 107.
[153] Settia, *pratica e teoria*, 96.
[154] Theodore Palaiologos, *Enseignements*, 55.
[155] Theodore Palaiologos, *Enseignements*, 55, 84.
[156] W. Caferro, *John Hawkwood. An English Mercenary in Fourteenth-Century Italy* (Baltimore, 2006), 45.

of financial resources and not their ineffectiveness on the battlefield. Being a supporter of Andronikos II, Gregoras is critical of the Cretan and Alan mercenaries, whom he sees as ineffective and ill-disciplined mercenaries, motivated by greed. Similarly, for Gregoras while the decision to hire the Catalan Company was correct, Fortune brought disaster.

That the authors of the period criticised a particular group of mercenaries is not enough to prove that they opposed mercenary service, since they approved the employment of other groups of mercenaries. For instance, Pachymeres is critical of the attitude of the Catalans. However, he shows an understanding of the Cretans and the Alans. Gregoras admires the Catalan Company, but he is rather critical of the Cretan and Alan mercenaries. It should not be forgotten that throughout their accounts Pachymeres and Gregoras say nothing against the employment of Turkish and Cuman mercenaries, who made a substantial part of the Byzantine army at least until the end of the thirteenth century. In addition, both Gregoras and Akropolites emphasise the competence, strength and effectiveness of the 800 Western European heavy cavalrymen Theodore I hired and used against the Seljuks in Antioch in 1211. Akropolites, stressing the high quality of those 800 Latin cavalry soldiers, states that when the Latin emperor in Constantinople, Henry of Flanders was informed that these soldiers were annihilated, he exclaimed that Laskaris was defeated and not victorious.[157]

Thomas Magistros and Gemistos Plethon seem to reject the very idea of employing foreign mercenaries. Their objection reflects mainly their ideas about taxation and the classical works that influenced them. Being an emperor of the Romans and a member of the aristocratic elite, which monopolised military commands and considered participation in war as being a social obligation, Kantakouzenos expresses negative views about serving exclusively for pay. However, as an experienced general he was aware that battlefield realities, the logistical needs of war and developments in the art of war necessitated the employment of professional soldiers who were constantly mobile and possessed specialised technological knowledge.

[157] Akropolites, I, 27.

CHAPTER FIVE

CAMPAIGNS

Late Byzantine military activities were concentrated in western Asia Minor (until the early fourteenth century), Thrace, Macedonia, Epiros and the Peloponnese. Therefore, war-making had to adjust to the geography of the Balkans, which presents a very rugged and fragmented landscape dominated by mountains, which comprise two thirds of the surface area. This geographical background was made for tough campaigning conditions, particularly during winter.[1]

Byzantine military leaders were aware that spring was the ideal period to launch an expedition. Michael IX set out from Constantinople on 22 April 1302 to march on the Turks in Magnesia. During the civil war between Andronikos II and Andronikos III the latter ordered the troops from the provinces loyal to him to assemble in Didymoteichon on 15 March 1322. Planning to campaign against the Latins in Patras, Constantine Palaiologos, issued orders to everyone in the region of Androusa to come on 15 March 1429 with their weapons and most of their troops.[2] The spring campaigns meant that there were plenty of crops available for purchase or plunder as well as fodder, which was necessary for the feeding of the war horses. This implies that potentially the army would find a large amount of supplies at relatively low prices in the local markets during its march.[3] Winter was the period of preparations. In 1341, Kantakouzenos suggested that the Byzantine forces should campaign in the Peloponnese the following spring and therefore, a fleet should be constructed during the winter for the transport of troops

[1] See M. Hendy, *Studies in the Byzantine Monetary Economy* (Cambridge 1985), 21–25; J. Haldon, *The Palgrave Atlas of Byzantine History* (New York, 2005), 3.

[2] Pachymeres, IV, 341; Kantakouzenos,I, 136; Sphrantzes, 42. Further examples: In 1262, after their defeat at the hands of the army of the Latin principality of Achaia the Byzantines reassembled the following March: *Chronicle of the Morea*, vv. 5020–5027. In March 1338, the army under the command of Andronikos III campaigned in Epiros: *Kleinchroniken*, I, 80. The view that early spring was the ideal period for launching campaigns is expressed by the anonymous author of the *Story of Belissarios*. The fleet of the heroic general Belissarios set sail on 15 March: *Ιστορία του Βελισσαρίου*, 152.

[3] D.N. Engels, *Alexander the Great and the Logistics of the Macedonian Army* (Los Angeles, 1978), 27.

and supplies. In the meantime Kantakouzenos, leading the mercenaries and the more powerful of the *pronoia* holders, obviously those wealthy enough to support themselves in a long winter campaign, would campaign against the Albanians. He allowed the rest of the troops to stay at their homes and to make the proper preparations for the following spring campaign. Kantakouzenos' aim was to attack the Albanians and return to Constantinople by Christmas in order to begin the preparations for the campaign in the Peloponnese because as he writes, 'it is not proper for the army to stay in a hostile land during the winter.'[4]

Akropolites indicates that, John III and Theodore II were the exceptions to the rule and did not make any distinction between winter and spring/summer and they did not consider the weather conditions an impediment to their military plans. In his funeral oration for John III, Akropolites summarises the difficulties the Byzantine troops were having in winter campaigns. He praises John III for campaigning, 'in rain and snow and when the water is frozen and the farmers and shepherds find safety in their houses and cease to work.'[5] For instance, after his victory at Poimamenon (1223/1224), John III conducted long, though successful sieges, which stretched to the following winter, against cities held by the Latins.[6] Theodore II followed similar tactics and his army faced the difficulties caused by out of season campaigns. After seizing Veroe, Theodore II was forced to cancel his plans for further attacks because of the heavy snowfall, although Veroe provided the army with supplies. Theodore II was forced to stay inactive for six days and realising that there was nothing he could do and being aware that it is perilous to stay for long in a hostile area, he retreated to Adrianople after plundering everything around Veroe.[7] Heading from Adrianople to Tzepaina, the army of Theodore II was caught by strong winds and heavy snowfall at Makrolivada, which according to Akropolites, was located four *stathmoi* away from Adrianople. Akropolites states that, what caused dissatisfaction among the soldiers was the expected lack of supplies.[8] Similarly, summer campaigns imposed constraints. During

---

[4] Kantakouzenos, II, 81–2.
[5] Akropolites, II, 17.
[6] Akropolites, I, 36.
[7] Akropolites, I, 113.
[8] Akropolites, I, 120. For further examples see Kantakouzenos, I, 134, II, 187–188. In 1321, Andronikos III encamped for three days in Limpidarios, close to Kosmidion (1321). However, it was winter and due to the heavy rain he was forced to retreat, after two soldiers had died as result of the weather conditions. In the winter of 1341, Kantakouzenos, having established himself and his troops in Didymoteichon, was planning

his second campaign on the Sangarios (summer 1281), Michael VIII sent letters to his wife, his mother in-law and the patriarch, mentioning the difficulties caused by the lack of water.[9] In another example indicating that summer campaigns were subject to the constraints imposed by the unpredictability of weather conditions, in June 1342 Kantakouzenos was forced to stop his march on Thessalonica due to a sudden rain which made the Axios impassable. The delay resulted in the reduction of the supplies of his troops, forcing Kantakouzenos to withdraw to Gynaikokastro to collect new ones.[10]

As far as the administrative context of the campaigns is concerned, it seems that the process of supporting campaigning armies had been simplified. There is no information indicating the existence of officials comparable to the middle Byzantine provincial *protonotarioi*, who were informed in advance about the needs of the coming armies, and were in charge of co-ordinating them with the respective departments of the central government.[11] The sophisticated and elaborate system of supplying imperial expeditions, where each province was to supply the campaigning army with a specific amount of various kinds of food supplies and pack animals as described by the tenth-century treatises of Constantine VII (912–959), was not in effect in late Byzantium.[12] This is the inevitable result of the limited resources available to the late Byzantine state, of the small size of the campaigning armies and of the limited geographical scope of their expeditions. In his investigation of the size of late Byzantine armies Bartusis concludes that the largest campaign forces the Byzantines were able to assemble were no more than a few thousand.[13] The favourite figure of late Byzantine authors is 2,000.[14] The campaigning armies of the small states established in Frankish Greece were composed of a few hundred

to march the following spring on Constantinople. After being invited by the powerful men of Adrianople to seize the city he decided to march there. Trying to keep this operation secret he decided to pretend that he was campaigning against other Thracian cities. However, when he reached Maritza the river was frozen. The ice did not allow the use of horses to cross the river. Also the soldiers could not use small boats to cross the river. Consequently, Kantakouzenos was forced to stay inactive for twelve days waiting for the river to become passable.

[9] Pachymeres, II, 635.

[10] Kantakouzenos, II, 242–243.

[11] For the *protonotarioi* see Haldon, *Warfare State and Society*, 143–144.

[12] Constantine Porphyrogenitus, *Three Treatises*, especially text C.

[13] Bartusis, *The Late Byzantine Army*, 258–270.

[14] Akropolites, I, 27; Pachymeres, IV, 367; Kantakouzenos, I, 259; Gregoras, I, 317, 433, II 614, 628.

soldiers. For instance, according to a letter sent by James, the bishop of Argos, to Cardinal Angelo Acciaiuioli in 1385, the duke of Athens, Nerio Acciaiuioli had in his employ 70 lances, 800 Albanian cavalrymen and many infantrymen. The despot of the Morea, Theodore I Palaiologos, had roughly 200 cavalry and many infantrymen as well as Turkish mercenaries. Their common enemy, the Navarrese Company, could muster a force of up to 1,300 horsemen.[15] Moreover, the late Byzantine commanders could not plan in advance the precise quantity and nature of supplies the army would need, since most of the campaigns were supported by troops enlisted along the way according to emerging needs. For instance, the *Chonicle of the Morea* mentions that in 1262, the Byzantine general who was send from Constantinople against the Frankish principality in the Peloponnese, Makrenos, had in his command massive infantry forces, which he recruited locally by taking onto his side the local Slavs.[16] In 1273, campaigning against Thessaly John Palaiologos recruited troops along the way from Thrace and Macedonia.[17]

The treatise of Pseudo-Kodinos is the only source which links an official with the supplying of military expeditions. It relates that the *megas stratopedarches* was in charge of provisioning the army with foodstuff, drinks and whatever was needed.[18] Although the middle Byzantine *megaloi stratopedarchai* were high-ranking officers, who were associated with the *domestikoi* of the east and the west, there is no evidence to suggest that they were linked to the logistical support of the army.[19] Similarly, none of the known late Byzantine *megaloi stratopedarchai* is linked to the logistical support of campaigning armies.[20] This does not necessarily mean that Pseudo-Kodinos' testimony is a fabrication. Reliance on circumstantial prosopographical details to reconstruct the organisation of late Byzantine armies and the functions military officers performed or did not perform entails the

---

[15] y Lluch, *Diplomatarii*, 611–613; Setton, *Catalan Domination*, 175; Zakythinos, *Le despotat*, I, 148; Bonn, *Morée Franque*, 267. See also the discussion of E. Sygkelou, *Ο πόλεμος στο δυτικό ελλαδικό χώρο κατά τον ύστερο μεσαίωνα* [*War in Western Greece in the Late Middle Ages*] (Athens, 2008), 116–125.

[16] *Chronicle of the Morea*, 4655–4663.

[17] Gregoras, I, 111.

[18] Pseudo-Kodinos, 174.

[19] Haldon, *Warfare State and Society*, 118; N. Oikonomides, *Les listes de préséance byzantines des IXe et Xe siècles* (Paris, 1972), 334–335.

[20] For the identity of the late Byzantine *megaloi stratopedarchai* who were military commanders see Panagiotides, "Η επιμέλεια του στρατού," 292–297.

danger of misunderstanding the sources and producing overgeneralisations. For instance, Pachymeres mentions that the *megas stratopedarches* John Synadenos participated in the campaigns against Thessaly and the forces of Charles I of Anjou in Bellegrada in 1275/1276 and 1281 respectively and adds that he was one of the commanders of the Byzantine fleet when Andronikos II decided to reduce it in 1285.[21] Pachymeres does not analyse the function of the office of *megas stratopedarches*. He merely reports that its holder, John Synadenos, participated in the aforementioned military operations. Moreover, that Theodore II Laskaris claims that he reintroduced the office of *megas stratopedarches* shows that at some point before 1254 this office had ceased to exist.[22] The reintroduction of the office does not mean that the late Byzantine *megaloi stratopedarchai* performed the same functions with their predecessors. Titles could remain the same but functions change in accordance with the needs and context of the period. Consequently, it is possible that the late Byzantine *megaloi stratopedarchai* had different functions from their predecessors.

The most characteristic example concerning the arrangement of a campaign is provided by Kantakouzenos and refers to a military operation which was never undertaken. On the eve of the civil war between the two Andronikoi, in April 1321, Kantakouzenos had been ordered by Andronikos II to campaign in Thessaly. He tried to find reasonable justification to postpone this mission indefinitely, so that he would be able to stay close to his co-conspirators. He requested a competent military force and money for its maintenance. He likewise requested a sufficient amount of money for the hiring of mercenaries. When he was given 50,000 gold coins by the emperor, he claimed that one of the units, which was to campaign under his command, had just returned from another campaign and the soldiers needed some time to rest. Because it was Easter he encountered resistance in his attempts to convince the soldiers to join the army and leave their houses. Moreover, each individual soldier had to bring with him money, with which he would support himself during the campaign.[23] Beyond the particular circumstances that motivated Kantakouzenos' actions, this example provides rather useful information. It shows

[21] Pachymeres, II, 527, 685, III, 81.

[22] Theodore II, *Κοσμική Δήκωσις*, N. Festa (ed.), in *Giornale della società Asiatica Italiana* 11 (1897–1898), 97.

[23] Kantakouzenos, I, 87–88.

that the campaigning army was supplied directly by the central government, while the military commander had the authority to make specific requests concerning the composition of the troops and the amount of supplies. Furthermore, the individual soldier should cover part of his maintenance himself. It was also expected that the head of the campaign would hire mercenaries along the way.

Pachymeres' description of Michael Tarchaneiotes' expedition to Thessaly in 1284 provides a similar example. The emperor together with Michael Tarchaneiotes, who was to lead the campaign, were assembling forces. It seems that there was a dispute over the financing of the operation. Pachymeres reports that later the emperor claimed that he had asked the operation to be funded by the treasury but the council around him thought that it should be financed through public contributions. As result, a 10% levy was imposed over the *pronoiai* to finance this specific military undertaking. Michael Tarchaneiotes received the money and departed with a large force.[24]

Kantakouzenos' and Pachymeres' examples indicate that most likely no fodder, food supplies and money was collected at the point of departure. Rather, money was available to the commanders and their troops at the point of their departure, while at least the wealthier of the soldiers were expected to cover part of their maintenance through their own expenses. It is worth noting that in his account of a Catalan victory over the Byzantines, Ramon Muntaner comments that, the victors captured large booty because the Byzantine soldiers had brought money with them.[25] These examples also suggest that the baggage train of the Palaiologan armies was rather small. The presence of baggage trains is attested in the campaigns of the Nicaean armies. Theodore II ordered the assembly of carts from Macedonia and Thrace to carry war engines and food supplies for the troops for the campaign against the Bulgarians in Tzepaina.[26] The existence of baggage train is also mentioned in the campaign of John III in Macedonia in 1252. To alleviate the problem of supplies the emperor arranged for camels and mules to carry supplies from Veroia to Ostrovos (around 50 kilometres), where he had set up his camp.[27] The sources make no mention to the pres-

[24] Pachymeres, III, 81. Michael Tarchaneiotes was the son of Michael VIII's sister Martha. See *PLP*, 27505.

[25] Muntaner, II, 456.

[26] Akropolites, I, 120.

[27] Akropolites, I, 92.

ence and size of the baggage train used by the campaigning armies of the Palaiologoi.

Supplies of food and fodder were provided through the imposition of the *mitaton*, an army levy in the form of purchases of provisions at nominal prices or requisition of supplies.[28] The *mitaton* shows how the money provided to campaigning armies at the point of their departure was used. The *mitaton* is known through the surviving exemptions granted by emperors to monasteries and individual private landowners. Usually the exemption grants which refer to a *mitaton* deal with local officials and local governors (*kephalai*) who, however, were also military commanders. Nonetheless, it is certain that this tax was demanded for the benefit of the army on the move. That the *mitaton* involved the sale of food supplies in distorted prices is evident in the *chrysobull*, which defines the privileges Andronikos II granted the city of Ioannina. According to this document

> There shall be no *mitaton* on the said properties, nor shall there be any other imposition on them, but those possessing these properties shall sell their products as they are sold in the area, and will not be forced at all to sell them contrary to the custom.[29]

Another example is provided by an exemption granted to the monastery of Zographou in 1325. The possessions of the monastery are exempted from every tax, as well as from the *mitaton* of the present and future *kephalai* found there, and from the *mitaton* of the army.[30] In 1332, Andronikos III issued a *chrysobull* prohibiting the governors of Serres from imposing the *mitaton* on the *paroikoi* on the lands of the Prodromos monastery, either for their own advantage or for the advantage of the soldiers.[31]

There is evidence showing that the *mitaton* was open to abuse. Michael VIII issued a *prostagma*, which orders the *kephalai* and their subordinates:

---

[28] Maksimović defines the *mitaton* as a burden on the local population obligating them to make provisions available for sale to officials and troops at a price below the prevailing market price: Maksimović, *Provincial Administration*, 157–159.

[29] MM, V, 82.

[30] *Actes de Zographou I. Actes Grecs*, W. Regel - E. Kurtz. - B. Korablev (eds.), (St. Petersburg, 1907), 44–46; M. Bartusis, "State Demands for the Billeting of Soldiers in Late Byzantium," *ZRVI* 26 (1987), 118.

[31] *Les archives de Saint-Jean Prodrome*, 66.

> Not to take anything without paying, whatever it may be, even an egg or a fruit, nor yet, by appearing to purchase the requisite product to harm the vendor so as to give him less than the proper price, one piece instead of two, and two *hyperpyra* instead of four.

The heads of armies are ordered to make sure that merchandise is sold fairly. The commanders of the armies should allow neither the vendors to sell just as it pleases them and harm the buyers, nor yet the purchasers to grab beyond just measure or buy edibles and drinkables in excess of what they need and then sell them for shameful profit.[32]

The *chrysobull* of Andronikos II in favour of the see of Kanina in Albania, which was issued in 1307, provides us with another example of abuse of the *mitaton*. The bishop of Kanina had requested that the possessions of his see be exempted from the obligation of furnishing grain for the *mitaton* of the regional heads. They pay for it but in much lower prices. In response to this request Andronikos II granted the see of Kanina exemption from the *mitaton*.[33]

The above-mentioned Kantakouzenos' and Pachymeres' accounts, as well as Michael VIII's *prostagma* show the important role of the commander in supplying campaigning armies. This is also illustrated in his ability and authority to set up camp in places and cities that could supply the army with food and fodder. Campaigning against the Bulgarians, Theodore II set up camp in Vatkounion, where he found plenty of supplies that could support the army for many days.[34] In 1256, Michael Palaiologos and Michael Laskaris set up camp in Edessa (Vodena) in a place, which would provide them with sufficient fodder.[35] John III (1246), and later Kantakouzenos (1342), encamped for a few days in Vera, a place that provided abundant supplies of fodder.[36] However, it seems that the presence of the army was not always welcomed by landowners. Privileged monasteries and lay landowners sought and received exemption from the obligation to permit the army to set up camp on their land. In the late Byzantine period this obliga-

---

[32] L. Burgman – P. Magdalino "Michael VIII on Maladministration. An Unpublished Novel of the Early Palaiologan Period," *FM* 6 (1984), 383, whose translation this is.

[33] P.J. Alexander, "A *chrysobull* of the emperor Andronikos II Palaiologos in Favor of the See of Kanina in Albania," *B* 15 (1949–1951), 181, 184–185.

[34] Akropolites, I, 121.

[35] Akropolites, I, 146.

[36] Akropolites, I, 72; Kantakouzenos, II, 196.

tion was called *aplekton* (camp), *pezeuma* (from *pezos* 'on foot'), *and pesimon* (from to fall').[37]

Markets were another source of supply. In 1246, John III on his way back from Melnik set up camp in Thessalonica and asked its head, the despot Demetrios Angelos, to arrange a market outside the city, so the soldiers would be able to buy the necessities.[38] In 1343, Kantakouzenos' troops were taken by surprise by the brigand Momčilo, while they were buying supplies in the market of Koumoutzena (Komotini).[39] The above-mentioned examples indicate that the armies could either use the existing markets of the cities or special markets were arranged for the particular occasion of the passing of an army. In addition, it seems safe to conclude that the army camp was followed by merchants, particularly during a siege when the troops would stay for a relatively long period of time in a specific place.[40]

The sources indicate the existence of storehouses, which would supply passing armies. However, they do not provide information concerning their role and nature, and therefore it is not clear whether their primary role was to supply troops stationed in the cities and provinces or campaigning armies passing through the area. The first instance connecting storehouses with the supply of military forces are the measures taken by John III for the defence of Asia Minor against a possible Tatar onslaught. John III fortified the fortresses with arms and foodstuffs, which would be stored properly so the fortresses would be able to withstand long sieges. He also made arrangements for foodstuffs to be transported from the countryside into the fortresses, while he ordered the registration of weapons assigning specialist artisans to construct the required number of arms.[41] However, this information refers to urgent measures aimed at defence against the Mongols, an enemy about whose strength and intentions the Byzantines, until then, had little knowledge.[42] It seems that in this case the role of the storehouses

[37] For the exemptions from *aplekton* and the relevant sources see Bartusis, "State Demands," 121–123. For the earlier periods see T. Kollias, "Essgewohnheiten und Verpflegung im Byzantinischen Heer," in W. Hörandner *et al.* (eds.), *Byzantios*. Festschrift für Herbert Hunger zum 70. Geburtstag," (Vienna, 1984), 193–202.

[38] Akropolites, I, 82.

[39] Kantakouzenos, II, 430; Gregoras, II, 706.

[40] Kananos (9) states that in the siege of Constantinople in 1422 the Ottoman camp was followed by many artisans and traders.

[41] Pachymeres, I, 187; *Synopsis Chronike*, 506–507.

[42] Pachymeres, I, 187, 'In that time this nation was completely unknown, so some said that they had the heads of dogs and they were eating humans.'

was to provide supplies both for the civilian population and the local troops. In another example, describing the advance of Michael II, the ruler of Epiros in Macedonia in 1257, Akropolites reports that the defenders of Prilep retreated to the granary.[43] According to Gregoras, at the outbreak of the civil war, in 1321, the properties of the supporters of Andronikos III were confiscated and given to the soldiers loyal to Andronikos II. Moreover, the storehouses of grain and wine were plundered.[44] It is worth noting that Thomas Magistros counsels the emperor to introduce storehouses in cities to be used in case of sieges.[45]

Booty covered much of the maintenance cost of marching armies.[46] The absence of an elaborate system of logistical support and the severe financial difficulties the late Byzantine state experienced meant that plunder provided the army with essential supplies. Rulers and military commanders had to provide their soldiers with the opportunity to loot to ensure their subsistence. Long before the thirteenth century the Byzantine generals knew that raids were the cheapest and safest way of undermining the strength of the enemy and of rewarding and feeding Byzantine soldiers.[47] Throughout late medieval Europe there were established customs and laws which regulated the division of booty. These rules varied according to the customs of different states. Since the ruler ran the greater risk in war, he was to receive the largest portion paid from of the captured booty. For instance, the royal portion in England was the one third, in Spain one fifth, in France the same portion of any booty exceeding 10,000 francs.[48] As far as late Byzantium is concerned, describing Andronikos III's campaign against the Albanians in Epiros in 1338, Kantakouzenos states that it was customary, however much booty the army seized that the one

[43] Akropolites, I, 149–150.

[44] Gregoras, I, 397.

[45] *PG* 145, col. 484.

[46] For a detailed discussion about the role and distribution of booty in late Byzantine warfare see S. Kyriakidis, "The Division of Booty in Late Byzantium. 1204–1453," *JÖB* 58 (2009), 164–176.

[47] For relevant examples see the *Taktika*, IX. 21–22; McGeer, *Sowing the Dragon's Teeth*, 147. For the impact which the prospect of booty had on the motivation of soldiers in the middle Byzantine period see Dagron – Mihaescu, *Le traité sur la guerrilla*, 231–234; A. Dain, "La partage du butin de guerre d' après les traités juridiques et militaires," in *Actes du VIe congrès d' études byzantines*, I, (Paris, 1950), 347–352.

[48] See the discussion of M. Keen, *The Laws of War in the Late Middle Ages* (London, 1965), 146.

fifth (*pempte moira*) should be given to the emperor as a reward and an equal part to the *megas domestikos* because he was the head of the whole army.[49] However, Kantakouzenos does not specify how the rest of the booty should be divided. Pseudo-Kodinos writes that,

> From the booty (*koursos*), first a fifth (*pentamoiria*) is given in the name of the emperor, a second portion on behalf of the entire army to the *megas domestikos*, and a third to the division leaders, from the division each commands.[50]

Kantakouzenos and Pseudo-Kodinos are the only authors of the period who specify that the spoils should be divided into fifths. However, they do not mention when this custom was established. To answer this question, it is necessary to compare their testimonies to the evidence provided by sources from earlier periods. The Byzantine legal texts and collections of secular law including the *Hexabiblos*—which was compiled by Constantine Harmenopoulos in 1345 but consisted mostly of excerpts of the tenth-century *Basilika*, themselves a digest of Justinianic law—unanimously state that one sixth of the booty collected in war should be given to the state (*demosion*). The rest of the loot should be equally divided among the officers and the soldiers.[51]

Although they emphasise the importance of the orderly and fair distribution of spoils, the military treatises of the early and middle Byzantine periods do not generally provide any details about the division of booty. The only exceptions are the *Sylloge Tacticorum*, which was compiled in the first half of the tenth century, and the *Taktika* of Leo VI.[52] Being influenced by older military manuals and law codes, the author of the *Sylloge Tacticorun* states that one sixth of the booty was given to the state and the rest was shared equally among the soldiers and their officers. He also reports that in the past the state received one tenth of the booty.[53] According to the *Taktika*, the commander of the army should divide the booty according to the imperial

---

[49] Kantakuzenos, I, 498.

[50] Pseudo-Kodinos, 251.

[51] *JGR*, II, 88 (Leges militares), II, 227 (Procheiros nomos), VI, 213, (Ecloga ad procheiron mutata), VII, 312 (Procheiron auctum). On the dating of the *Hexabiblos* see M.T. Fögen, "Die Scholien zur Hexabiblos im Codex vetustissimus Vaticanus Ottobonianus," *FM* 4 (1981), 268–275.

[52] For the dating of *Sylloge Tacticorum* see Dain, "Les stratégistes Byzantins," 357–358; McGeer, *Sowing the Dragon's Teeth*, 184; Dragon - Mihaescu, *Le traité sur la Guérilla*, 154.

[53] *Sylloge Tacticorum*, 99; Dain - Foucault, "Les stratégistes Byzantins," 351.

laws, so one fifth of what the army captures should be given to the state (*demosion*) and the rest equally shared among the officers and the common soldiers.[54]

Consequently, Pseudo-Kodinos and Kantakuzenos, who specifies that he refers to an old custom, describe a practice which was in effect when the *Taktika* were compiled. The lack of sources prevents us from reaching any conclusion about what led to the increase in the share of the state. It is worth noting that the *Taktika* include numerous references to the conflict between the Byzantines and the Arabs and Islamic tradition specifies that one fifth of the booty seized in war should be reserved for the state.[55] Does this mean that the demand of the state for one fifth of the booty shows an Arab-Islamic influence?[56] Probably, yes. The possibility of an Islamic influence is supported by the argument that the imposition of a one-fifth tax on the booty which the Christian municipalities captured during the conflict between the Christians and the Muslims in Spain was a tax derived from the Muslim practice of allocating one fifth of the booty to the state.[57] This seems to be confirmed by Muntaner, who states that when the Catalan Grand Company established its basis in Gallipoli he was put in charge of collecting the one-fifth of the booty captured in the raids carried out by the Company.[58]

Unlike the earlier sources, which state that the soldiers and the officers received equal shares, Kantakouzenos and Pseudo-Kodinos relate that the *megas domestikos* and the division commanders should receive a larger share than their soldiers. This can be seen as a result

---

[54] *Taktika*, XX. 192.

[55] For the references to Islam in the *Taktika* see G. Dragon, "Byzance et le modèle islamique au Xe siècle, à propos des Constitutions tactiques de l' empereur Léon VI," in *Comptes rendus des séances de l' Académie des Inscriptions et Belles-Lettres* (Paris, 1983), 219–243; T. Kolias, "The *Taktika* of Leo VI the wise and the Arabs," *Graeco-Arabica* 3 (1984), 129–135. For the division of the spoils of war according to the Islamic tradition see M. Khaddouri, *War and Peace in the Laws of Islam* (Baltimore 1955), 120–123; N. Calder, "Khums in Imami Shi'I Jurisprudence, from the Tenth to the Sixteenth Century A.D.," *Bulletin of the School of Oriental and African Studies*. 45 (1982), 39–47; A. Sachedina, "Al-Khums: The Fifth in the Imami Shi 'I Legal System," *Journal of Near Eastern Studies* 39 (1980), 276–277.

[56] For the warfare between the Arabs and the Byzantines see J. Haldon – H. Kennedy, "The Arab-Byzantine Frontier in the Eighth and Ninth Centuries: Military Organisation and Society in the Borderlands," *ZRVI* 19 (1980), 79–116; McGeer, *Sowing the Dragon's Teeth*, 225–248.

[57] Powers, *A Society Organized for War*, 163.

[58] Muntaner, II, 465.

of the dwindling resources of the Byzantine state. The state found it difficult to pay and reward its officers through cash payments. Instead, it was more convenient for the government to provide its officials with sources of wealth. Spoils were such a source and in a period of endemic warfare booty could provide ample opportunities for enrichment. Moreover, in Western Europe generals received a customary portion of the booty. For instance in England they could claim one third of the spoils (which was equal to the royal portion), one tenth in France, and one seventh or one tenth in Spain.[59]

Furthermore, Pseudo-Kodinos observes that, the commander of the *kourseuontes*, the *protostrator*, received a share of the booty captured in their raids. As he writes,

> The *protostrator* is the defender of the *kourseuontes*. The *kourseuontes* have neither rank-order nor standards. Instead, they are launched without order and the *protostrator* being behind them protects them if they get involved in fighting; for this he is called *protostrator*, because he is in advance of the entire army. And it is customary for the *protostrator* to receive from the variety of the captured animals those that are called *phytilia*.[60]

Pseudo-Kodinos' description of the *protostrator* is not in agreement with the function of this office in the middle Byzantine period.[61] This does not mean that Pseudo-Kodinos' testimony is a fabrication, rather that the function of the *protostratores* changed. Similarly, the term *kourseuontes* to describe the advance scouts or raiders derives from the *koursores* and *prokoursatores* of the earlier periods. However, their function was different.[62]

It is worth noting that the presence of troops who performed identical tasks to the *kourseuontes* mentioned by Pseudo-Kodinos is attested in non-Byzantine sources which are chronologically close to the treatise of Pseudo-Kodinos and the *History* of Kantakouzenos and

---

[59] Keen, *Laws of War*, 146.

[60] Pseudo-Kodinos, 173.

[61] From the middle of the eighth century the *protostrator* was the head of the imperial esquires or mounted attendants of the emperor: Oikonomides, *Les listes de préséance*, 337–38; Constantine Porphyrogennitus, *Three Treatises* 253; Hohlweg, *Beiträge*, 59, 111–117.

[62] The *koursores* or *koursatoroi* were marching ahead of the rest of the army and their aim was to pursue the retreating enemy, engage the enemy in skirmishes or lay ambushes: McGeer, *Sowing the Dragon's Teeth*, 22, 26, 99, 182, 211; *Sylloge Tacticorum*, 63.

describe warfare in Frankish Greece and the late medieval Balkans. The *Chronicle of the Morea* relates that the *koursatoroi* were a unit of the army of the Latins of the Peloponnese. Their task was to advance one day ahead of the rest of the army.[63] Wishing to emphasise the effectiveness of the raids Carlo Tocco conducted on the possessions of the Albanian chieftain, Sguros Bua, the chronicle of Toccos comments that the *rymparika* captured prisoners 'in the manner of the Turks.' Marching against the Albanians in Angelokastron, Carlo Tocco ordered the *koursatoroi* to march ahead of the rest of his army and scout and plunder the area around the fortress.[64] The *asapi*, who are mentioned in a Venetian document in 1429 as a loyal group of soldiers, seem to have been similar to the *kourseuontes*. According to this document, two *asapi*, who were based in Thessalonica, Andreas and Theodore Olbofaci, appealed to the Senate to complain about the extra charges they had been held liable for in addition to the *pendamerea* which was the only tax they were obliged to pay.[65] Like the *pentamoiria* the *pendamerea* was a tax of the one fifth of the booty captured by the *asapi*. One of the two *asapi* is called 'the *protostrator*.' Consequently, it is possible that he was the head of the *asapi* raiders. It is likely that these *asapi* were similar to the Ottoman *azabs* who were light infantrymen recruited *ad hoc* for the expeditions of the Ottoman army.[66]

With regard to the distribution of booty, Pseudo-Kodinos relates that the *krites tou fossatou* (judge of the army) examines and resolves the differences between the soldiers concerning horses, weapons and booty.[67] According to Haldon, the origins of this office can be found in the middle Byzantine period and its establishment is linked to the changing character of the Byzantine army in the later tenth century and specifically to the employment of large numbers of foreign soldiers who fought under their own commanders. This development made necessary the appointment of an official to resolve the disputes between soldiers who came from diverse cultural background.[68] However, it has

---

[63] *Chronicle of the Morea*, vv. 3670, 6652.

[64] *Cronaca dei Tocco*, vv. 962–967, 227–230.

[65] Thiriet, *Régestes*, II, 258–259.

[66] C. Imber, *The Ottoman Empire 1300–1600. The Structure of Power* (London, 2002), 259–260.

[67] Pseudo-Kodinos, 184.

[68] J. Haldon, "*The krites tou stratopedou*: a New Office for a New Stuation?' in C. Morrison (ed.), *Mélanges Gilbert Dagron*," *TM* 14 (2002), 279–286.

been argued that there is insufficient information to link the *krites tou stratopedou* or *tou fossatou* with the rise of the number of foreign mercenaries and it is unclear whether the *krites tou stratopedou* was an official title. Relying on the fact that the few surviving references do not connect the known *kritai tou fossatou* with military operations Goutziokostas concludes that the *krites tou fossatou* was a title without any content.[69] However, it is impossible to find specific information about the function of offices relying on limited circumstantial and prosopographical details provided by historical accounts and documents. It is also worth noting that from the late fourteenth century, the officer who at least theoretically was in charge of executing the delivery of the one fifth of the booty seized by the Ottoman armies to the sultan was the judge of the army.[70] If this is a Byzantine influence on the Ottoman military administration, it confirms Pseudo-Kodinos' statement about the function of the judge of the army.

In wars against co-religious enemies booty consisted of movable goods and not of prisoners. In the late medieval West the rule under which prisoners were reduced to slavery did not apply in wars between Christians.[71] According to Kantakouzenos the armies of the Christian Orthodox powers were prohibited from enslaving the defeated soldiers and civilians of a co-religious enemy. Describing a large-scale raid of Andronikos III on Bulgaria in 1322, Kantakouzenos reports that the emperor plundered the land for thirty days and seized livestock and other property. However, he did not capture any prisoners because it was customary for the Byzantines and the Bulgarians not to enslave each other.[72] Similarly, in the account of the battle of Velbužd, which was fought in 1330, he relates that the victorious Serbian army under the command of Stefan Dečanski (1321–1331) stripped the defeated Bulgarian soldiers of the army of Michael Šišman of their weapons and released them.[73] Kantakouzenos adds that when Andronikos III campaigned against the Albanians in Epiros in 1338, the Byzantine army captured a huge number of oxen, horses and sheep. However,

---

[69] A. Goutziokostas, "Ο κριτής του στρατοπέδου και ο κριτής του φοσσάτου," [The *krites tou stratopedou* and the *krites tou phossatou*] *Byzantina* 26 (2006), 87–90.

[70] I. Beldieanu-Steinherr, "En marge d'un acte concernant le penğyek et les aqinği," *REI* 38 (1969), 39–41; C. Kafadar, *Between Two Worlds. The Construction of the Ottoman State* (Los Angeles 1995), 142.

[71] Keen, *Laws of War*, 137.

[72] Kantakouzenos, I, 171.

[73] Kantakouzenos, I, 429.

unlike their 2,000 Turkish allies sent by Umur, the Byzantine soldiers did not enslave the Albanians because, as Kantakouzenos writes, they were not permitted to do so unless the enemies were barbarians who did not believe in Jesus Christ.[74] Kantakouzenos is confirmed by the fifteenth-century historian Doukas, who observes that it was an old custom neither to kill nor to enslave prisoners of the same religious faith. Instead, after having their equipment and supplies taken away, the soldiers of the defeated army were released.[75]

However, neither Kantakouzenos nor Doukas specify whether this rule applied to non-Orthodox Christian mercenaries employed by an Orthodox power. For instance, a substantial part of the Serbian army which fought in the battle of Velbužd was made up of German mercenaries. Furthermore, unlike the Western European practice, according to which Christian captives were ransomed, Kantakouzenos and Doukas seem to imply that the soldiers of co-religious enemies were released immediately after the end of the battle without having to pay any ransom.[76] When the custom arose of not enslaving the soldiers of a co-religious enemy is unknown. It is worth noting that the main source for the history of the so-called empire of Nicaea, George Akropolites, writes nothing about it. Instead, his account provides evidence which contradicts the reports of Kantakouzenos and Doukas. Akropolites relates that in 1255, during his campaign against the Bulgarians, Theodore II seized a significant amount of booty and many prisoners, including men, women and infants.[77] In 1257, in a battle against the Epirots, Michael Palaiologos captured Theodore Angelos, the bastard son of Michael II Angelos, ruler of the despotate of Epiros. According to Akropolites, Theodore pleaded with Michael Palaiologos to spare his life. However, being unable to identify him, Michael Palaiologos handed him over to a Turk who killed him.[78]

There are many examples showing that booty was used as an important means to feed and motivate Byzantine armies. The decision of the Nicaean rulers to exempt the frontier soldiers in Asia Minor from any levy on the spoils they seized demonstrates the importance of

[74] Kantakuzenos, I, 497.
[75] Doukas, 57, 99.
[76] For the treatment of Christian prisoners in Western Europe see Keen, *The Laws of War*, 137.
[77] Akropolites, I, 113.
[78] Akropolites, I, 148; Nicol, *The Despotate of Epiros*, 163–164.

booty in the continuous frontier warfare between the Nicaeans and the Turcoman chiefdoms. Pachymeres comments that one of the main reasons why the eastern frontier was well protected during the Nicaean period was the establishment of frontier guards who were granted numerous privileges and incentives not to abandon their lands and the immunity from any tax on the booty they captured as a result of their raids on enemy territory was one of them. The same author considers the abolishment of this privilege as one of the most important factors that led to the weakening of the Byzantine defence in Asia Minor. He concludes that this prompted many locals to join the enemy, hoping to earn more through plunder.[79] In 1263, complaining that they had not received their salaries for six months, the Turkish mercenaries employed by the Byzantines in the Peloponnese deserted their paymasters.[80] Shortly afterwards, they were hired by the Latins and it was agreed that they would keep for themselves all the booty that they captured.[81] This demonstrates that booty was an important part of the payment of the Turkish mercenaries. In 1344, after capturing Gratianou, Kantakouzenos paid his mercenaries from the booty which was seized there.[82]

Apart from cash, booty and the *mitaton* a significant number of examples indicate that at least the wealthier soldiers were expected to supply themselves through their own resources. As has been discussed in a different context, Pachymeres and Kantakouzenos connect the *pronoia* system with the ability of the soldiers to provide themselves with the necessary equipment and supplies when they are ordered to participate in military operations. The argument that the soldiers were expected to support themselves is shown by the fact that the army was followed by servants, who were employees of individual soldiers.[83] However, there is not any information specifying whether the emperor or the head of the campaign could impose a limit on the number of supplies and servants each soldier would bring along, nor is anything known concerning the amount of supplies these servants carried. The

[79] Pachymeres, I, 29–31, 35.

[80] *Chronicle of the Morea*, vv. 5099–5100; Sanudo, 126.

[81] *Chronicle of the Morea*, v. 5674.

[82] Kantakouzenos, II, 426.

[83] For the terms used by the sources to identify these people T. Kollias, "Ein zu wenig bekannter Faktor im byzantinischen Heer. Die Hilfkräfte," in C. Scholz, (ed.) *ΠΟΛΥΠΛΕΥΡΟΣ ΝΟΥΣ. Miscellanea für Peter Schreiner zu seinem 60. Geburtstag* (Munich/Leipzig, 2000), 113–124.

servants of the soldiers appear in four instances. In 1246, to solve the problem of supply shortages in the siege of Serres John III stirred up the servants of the soldiers, the so-called *tzouloukones*, to storm and pillage the lower part of the city, so they could get possession of plenty of supplies.[84] In another example, Akropolites mentions the presence of the servants of the soldiers in the campaign of Theodore II in Tzepaina in 1255.[85] In 1322, while the troops of Andronikos III besieged the fort of Sakkoi near Selymvria, the servants of the soldiers were enraged by the insults of its defenders, so they requested and were given permission to storm and seize the fort themselves.[86] Kantakouzenos' states that in 1343, when in a battle against Momčilo, somewhere between Mesene and Komotini, his horse was killed, one of his retainers, Lantzaretos, dismounted from his horse and gave it to him. In that battle Lantzaretos was seriously injured.[87] It is worth noting that in almost all of the above-mentioned cases the retainers participated in the actual fighting. However, Akropolites specifies that the *tzouloukones* were not armed and that they improvised weapons on the spot. This implies that they were not supposed to fight. Kantakouzenos calls Lantzaretos, an armed servant, an *oiketes*. This indicates that the servants who accompanied him were also a kind of armed private retinue.[88] Similarly, in 1342 in Thessalonica during the skirmishes between the zealots and the retreating troops of the *protostrator* Theodore Synadenos, Kantakouzenos observes that some of the *oiketai* of Synadenos were injured.[89] Bearing in mind the social status and wealth of people like Kantakouzenos and Synadenos, it would be expected that during campaigns they could be accompanied by their private armed servants, who would be fed and paid by them and not by the central government.[90]

The treatise of Pseudo-Kodinos is the only source of the period which is concerned with the details of setting up camp and providing for its security. Pseudo-Kodinos relates that the *epi tou stratou* is in charge of finding a place to set up camp. If the *megas domes-*

---

[84] Akropolites, I, 75.
[85] Akropolites, I, 120.
[86] Kantakouzenos, I, 144.
[87] Kantakouzenos, II, 433.
[88] For a discussion of the role of retainers and servants in the late Byzantine army, see Bartusis, *The Late Byzantine Army*, 221–234.
[89] Kantakouzenos, II, 235.
[90] For the *oiketai* see Weiss, *Kantakuzenos*, 139–148.

tikos does not approve the choice of the *epi tou stratou*, the army will set up camp in another location selected by the *megas domestikos*.[91] Unlike Pseudo-Kodinos, the military treatises of the earlier periods report that the so-called *minsouratores* or *mensores* were in charge of finding a suitable place to set up camp, while the task of the *antikensorai* was to find information about the water and fodder supplies the army could use.[92] Late Byzantine authors indicate that the armies set up camp in places which could provide the troops with sufficient supplies of food, water and fodder. However, they do not name officers and soldiers who went ahead to lay out the campsites.[93] This illustrates the circumstantial nature of the information provided by the historians of the period. Another factor that needs to be borne in mind is the limited geographical scope of the campaigns undertaken by late Byzantine armies which made the presence of such troops superfluous. In addition, that the office of *epi tou stratou* appeared for the first time in the late Byzantine period indicates that Pseudo-Kodinos' description of the function of this office relies on practices that were followed after 1204.[94]

According to Pseudo-Kodinos, following the *megas domestikos*' orders, the *megas drouggarios tes vigles* arranges the guard duty. Pseudo-Kodinos comments that, the garrisons are usually manned by archers during the day. However, archers should not be posted in night watch. When night approaches the *megas drouggarios tes vigles* orders the guards to get food and feed their horses. Then he sends them back to their watch duty. One of the duties of *megas drouggarios tes vigles* is to patrol and supervise the night watch.[95] The connection of the office of *megas drouggarios tes vigles* with the security of the military camp is confirmed by earlier sources. A tenth-century treatise on imperial expeditions relates that, one of the duties of the *drouggarios tes vigles* was to patrol the camp during the evenings and during the nights and

---

[91] Pseudo-Kodinos, 248–249.

[92] The term *minsouratores* is from the Latin *mensor*. It is used both of the surveyors who went ahead to choose a suitable location for the camp and mark out its boundaries and of the officials responsible for the disposition of the imperial tent on campaigns or other imperial outings: McGeer, *Sowing the Dragon's Teeth*, 51, 76; Haldon, *Warfare State and Society*, 160; *Strategikon*, I.3, II.12; *Taktika*, IV.24, XII.43; Dennis, *Treatises*, 84–85, 249–265.

[93] For examples see Akropolites, I, 72, 121, 146; Kantakouzenos, II, 196, 430; Gregoras, II, 706.

[94] For the office of *epi tou stratou see* Guilland, *Recherches*, I, 527–528.

[95] Pseudo-Kodinos, 249–251.

names him as the officer in charge of the camp security. The *droug-garios tes vigles* was also in charge of controlling who is going out and who is readmitted to the camp in accordance with agreed procedures, such as passwords.[96] Therefore, Pseudo-Kodinos reflects the continuation of an old practice. It is impossible to identify the source that led Pseudo-Kodinos to state that archers perform guard duty only during the day. The known military manuals rarely specify the type of soldiers who should guard the camp. For instance, the sixth-century anonymous treatise *On Strategy* focuses on the construction of the static defences of the camp and not on the type of soldiers who should man its garrison, while a tenth-century military manual relates that the watch posts should be manned by infantry, javelin throwers and archers; horsemen should stand guard only in those areas where it is suspected that the enemy will attack.[97]

Pseudo-Kodinos relates that when the *megas domestikos* is inspecting the army, the *megas admouniastes* follows him and arranges the provision of horses and arms to any soldiers who lack them.[98] The function of the *megas adnoumiastes* as described by Pseudo-Kodinos is similar to the accounts provided by the sources of the earlier epochs which portray the *adnoumiastes* as the officer in charge of the census, enumeration and inspection of the soldiers of the imperial armies. The muster roles (*adnoumia*) were kept at the *stratiotikon logothesion*.[99] The *Strategike Ekthesis* recommends the general to make a census (*adnoumion*) before and after the campaign and to maintain accurate records of the available manpower and equipment.[100] Pseudo-Kodinos does not connect the *adnoumion* to the *stratiotikon logothesion* which had already declined in the eleventh century.[101] Instead, he relates that the *logothetes to stratiotikou* has no function.[102] That the *megas adnoumiastes* followed the *megas domestikos*, also indicates that the latter had taken up the responsibilities of the *logothetes tou stratiotikou.*

---

96 Constantine Porphyrogenitus, *Three Treatises*, 121–122, 131.

97 Dennis, *Treatises*, 89–92, 265–267.

98 Pseudo-Kodinos, 250.

99 The office of *logothetes tou stratiotikou* appeared in the seventh century and its holder was in charge of the recruitment and financing of the army and of keeping the muster roles (*stratiotikoi kodikes*): Oikonomides, *Les listes de préséance*, 314.

100 Dennis, *Treatises*, 320–322.

101 Oikonomides, "Organisation," 135–136.

102 Pseudo-Kodinos, 184.

CHAPTER SIX

FORTIFICATIONS AND SIEGES

The role of fortifications

The endemic warfare between small states which lacked the resources and manpower to control large territories increased the importance of static defences. The defence of the principalities which were established in the Balkans in the late medieval period relied on the possession of heavily fortified cities and fortresses in strategic locations. This system of defence necessitated the repair and maintenance of existing walls and fortresses and favoured the construction of new, heavily fortified citadels and fortifications.[1] Much of the military activity during the later medieval period in the Balkans and Asia Minor revolved around fortifications which were used to observe the enemy, to hold back fugitives, to assemble for raids against enemy territories and to protect the population and soldiers in case of war or enemy raids.

Since warfare in Byzantium was a state issue, the building and repair of fortifications was an imperial prerogative through which the throne displayed its sovereignty.[2] The ideological importance of the maintaining and building fortifications and walls is reflected in imperial panegyrics. In his first imperial oration Metochites praises Andronikos II for repairing the existing fortifications and building new ones during his stay in Asia Minor between 1290 and 1293. He remarks that the reinforcement of the empire's defences proves the emperor's generosity (*genaiopsychia*).[3] Metochites adds that Andronikos II constructed towns (*polismata*) and fortresses (*phrouria*) close to the frontier and close to each other and points out that the emperor exploited the geographical features of Asia Minor, such as rivers, inaccessible places

---

[1] See the discussions of S. Ćurčić, "Architecture in the Age of Insecurity. An Introduction to Secular Architecture in the Balkans," in S. Ćurčić – E. Hadjitryphonos (eds.), *Secular Medieval Architecture in the Balkans, 1300–1500, and its Preservation* (Thessalonica, 1997), 19–51.

[2] E. Ivison, "Urban Renewal and Imperial Revival in Byzantium (730–1025)," *BF* 26 (2000), 7.

[3] Metochites, *Orations*, 198–200.

and mountains. The emperor, as Metochites continues, fortified the Byzantine frontier from the Black Sea to Sangarios and to Bithynia, as well as the Byzantine possessions in Lydia and around the Meander. Relying on Hellenistic literary models Metochites concludes that, due to the emperor's acts of wisdom and bravery, the Byzantine frontier in Bithynia looks like a very large city. Sangarios is its body and backbone, while the newly built towns between the rivers (Sangarios and Meander) are its towers (*pyrgomata*).[4] Metochites' description of the building project of Andronikos II in Asia Minor recalls the advice provided by a sixth-century military manual, the anonymous author of which comments that rivers and inaccessible places are suitable points for the building of forts which should not be located too much out in the open.[5] This coincidence does not prove that Metochites had read the sixth-century text. Nonetheless, it shows the familiarity of the later Byzantines with the basic principles of building and organising static defences and their awareness that the systematic use of natural features provides an advantage for defensive fortifications. Moreover, archaeological evidence indicates that Metochites' remarks about the activities of Andronikos II are not a fabrication aimed at enhancing the role of the emperor in the defence of Asia Minor. In their survey of Byzantine fortifications, Foss and Winfield have identified sites which date from the Palaiologan period. For instance, they conclude that sections of the fortresses around Nikomedia were rebuilt or repaired by Palaiologan emperors and a number of the fortresses in Bithynia and along the Sangarios have been dated to the reign of Andronikos II's father, Michael VIII.[6]

Assessing Andronikos III's reign, Kantakouzenos points out the importance of the former's rebuilding projects in Thrace and Macedonia. More specifically, Kantakouzenos relates that Andronikos III built the walls of the fortress of Gynaikokastro and added a high tower to the structure in order to reinforce its defence against siege engines. He reinforced the walls of Siderokastron, rebuilt the walls

[4] Metochites, *Orations*, 364–370. For the influence of Hellenistic authors on Metochites' description of the frontier see the comments of Polemis, 60–65.

[5] Dennis, *Treatises*, 29–33.

[6] C. Foss - D. Winfield, *Byzantine Fortifications. An Introduction* (Praetoria, 1986), 158–159; C. Foss, *Survey of Medieval Castles of Anatolia II. Nikomedia* (Ankara, 1996), 64; idem, "Byzantine Malagina and the Lower Sangarius," *Anatolian Studies* 40 (1990), 175–176; C. Giros, "Les fortifications médiévales," in *La Bithynie au Moyen âge*, B. Geyer - J. Lefort (eds.), (Paris, 2003), 222.

of Amphipolis and brought in settlers from other cities. In Thrace, Andronikos III rebuilt Anastasioupolis and renamed it Peritheorion. Kantakouzenos adds that Andronikos III reconstructed Arkadioupolis (modern *Lüleburgaz*) on a grand scale, after it had been largely destroyed in Tatar raids and he established new settlers in the city, some of whom were of noble birth. As was the case with Metochites, who attributed Andronikos II's building programme in Anatolia to the fulfillment of imperial duty, Kantakouzenos points out that the need to reinforce the empire's defences was not the only incentive of Andronikos III. The emperor was also motivated by the love of honour (*philotimia*).[7]

Imperial propaganda praises Manuel II for constructing the wall of Hexamilion, with the aim of protecting the Peloponnese from Ottoman raids in the fifteenth century. Manuel Chrysoloras praises Manuel II for restoring the wall of Hexamilion, despite the vehement resistance of rebellious local lords who threatened its builders. These local magnates were resentful of the fiscal burden imposed on them and fearful of their freedom of action. Manuel II is also praised for using building materials of the highest quality and exceptional building techniques to construct the wall. Chrysoloras concludes that even if the enemy managed to use the engines invented by Archimedes, they would not destroy it.[8]

The cost of building fortifications was primarily covered through a levy, the so-called *kastroktisia* (castle building).[9] Most likely this tax was originally a corvèe which existed before the tenth century and was levied upon the civilian population for labour and materials.[10] The large number of documents referring to *kastroktisia* in late Byzantium shows that it had been commuted for cash, while in the fifteenth century another levy, the so-called *phloriatikon* appears in the Peloponnese. Its aim was to fund the construction and repair of the

[7] Kantakouzenos, I, 542.

[8] *PP*, III, 242–244. In a letter addressed to the monks David and Damian in 1416, Manuel II discusses the opposition of local lords to the building of the Hexamilion: Manuel II, *Letters*, 213–216; *Mazaris' journey to Hades, or interviews with Dead Men about Certain Officials of the Imperial Court*, Barry J.N. et als (ed.) (Buffalo, 1975), 82–85.

[9] For a detailed discussion about *kastroktisia* see S. Troianos, "Kastroktisia," *Byzantina* 1 (1969), 39–57.

[10] Troianos, "Kastroktisia," 47–48.

wall of Hexamilion.[11] Both the *kastroktisia* and the *phloriatikon* were levies imposed by the state and not by any town authorities, as was the case in parts of Western Europe. The Byzantine cities remained under the control of the central government and did not evolve into independent communities.[12] The imposition of these taxes did not prevent the military authorities from occasionally forcing the local population to construct and repair city defences as well. In 1341, the people of Didymoteichon were forced to dig a moat around the lower city in order to reinforce its defences against the army of the regency of Constantinople.[13]

That oftentimes the taxes for the maintenance of fortifications were not included in the exemptions granted by the throne to private individuals, towns and monasteries illustrates the need for the continuous maintenance of fortifications and the fact that these levies were the most important, if not the single, source of funding for the maintenance and repair of the static defences of the empire. For instance, the *chrysobull* that Andronikos II issued in 1319 in order to define the privileges of Ioannina mentions that the inhabitants of this city were exempted from paying the *kastroktisia* for any *kastron* except for repairs needed in their own *kastron*.[14] In 1280, Michael VIII granted lands and exemptions to the *protovestiarites* Mourinos, apart from *kastroktisia*.[15] In 1295, Michael Gavrielopoulos exempted the people of Phanarion in Thessaly from the *kastroktisia* for fortresses other than their own.[16] In 1427, Plethon Gemistos was granted the rights of the village of Vrysis in the Peloponnese, apart from the taxes for maintenance of the wall of Hexamilion.[17] It is worth noting that these exemptions granted by Andronikos II and Gavrielopoulos indicate that it was not uncommon for the state to impose levies on a town for repairing and maintaining fortifications which were located in another territory.

Despite its ideological importance the building and repair of fortifications gradually stopped being the strict prerogative of the emperor.

---

[11] Troianos, "Kastroktisia," 53–54; Bartusis, "State Demands," 205, 207.

[12] A. Kazhdan, "The Italian and Late Byzantine City," *DOP* 49 (1995), 2; M. Wolfe, "Siege Warfare and the *Bonnes Villes* of France during the Hundred Years War," in I.A. Corfis - M. Wolfe (eds.), *The Medieval City under Siege* (Woodbridge, 1995), 58.

[13] Kantakouzenos, II, 289.

[14] MM, V, 82.

[15] *Actes de Docheiariou*, 108.

[16] MM, V, 260.

[17] MM, III, 173. For further examples see Bartusis, "State Demands," 206–209.

In the late thirteenth century, Constantine Palaiologos, the second son of Michael VIII, is one of the first known aristocrats to have built his own fortified tower in Nymphaion.[18] Under Andronikos III it was not uncommon for wealthy individuals to build fortifications, to fund imperial campaigns and to contribute to the payment of soldiers. John Kantakouzenos built the tower in Pythion and John Vatatzes the fortifications of Teiristasis in Thrace.[19] Alexios Apokaukos built the fortress in Epivatai.[20] It is probable that the severe financial difficulties of the empire after the civil war between the two Andronikoi led the emperor to allow wealthy individuals to construct fortifications in Thrace which could be used for the defence of the area against a possible large-scale Serbian attack and against the raids of the Turks. It is worth noting that people such as Kantakouzenos and Apokaukos were close associates of the emperor and not all imperial subjects were allowed to build private forts. Andronikos III campaigned against Martino Zaccaria, the Genoese governor of Chios, who was ruling the island recognising the nominal sovereignty of the Byzantine emperor, on the pretext that Zaccaria had started building a new fortress and fortifications without the emperor's permit.[21]

Nevertheless, the building of private fortifications was not something new in the history of the Byzantine empire. Long before our period, the emperor Theodosius II (405–450) had introduced a law which allowed private individuals to fortify estates provided they had the support of the local community. For instance, a fifth-century magnate who founded a fortified site justified his act by presenting himself as the protector of his community against forces hostile to the emperor Honorius.[22] It should not be forgotten that Kantakouzenos justified the money he gave to the imperial soldiers by claiming that he

---

[18] Pachymeres, III, 251.

[19] Kantakouzenos, II, 195, 475; Gregoras, II, 708. For a detailed examination of the tower of Pythion see K. Tsouris – A. Mprikas, *Το φρούριον του Πυθίου και το έργο της αποκατάστασης του. Προκαταρκτική Ανακοίνωση* [*The Fortress of Pythion and the Work for its Restoration. Preliminary Report*] (Kavala, 2002), passim; R. Ousterhout – Ch. Bakirtzis, *The Byzantine monuments of the Evros/Meriç River Valley* (Thessalonica, 2007), 145–159.

[20] Kantakouzenos, II, 475; Gregoras II, 585, 741–742.

[21] Kantakouzenos, I, 370–388; Ph. Argenti, *The Occupation of Chios by the Genoese*, 3 vols. (Cambridge, 1958), I, 55–62; M. Balard, *La Romanie génoise : XIIe-début du XVe siècle*, 2 vols. (Rome, 1978), I, 69–72; Bosch, *Andronikos III*, 112–118.

[22] See S. Connely, "Fortifying the City of God: Dardanus' Inscription Revisited," *Classical Journal* 102 (2007), 146.

wished to be outstripped by no one in giving money and servants, as well as his services, for the benefit of the people of the same ethnicity and for the honour of the emperors.[23]

Furthermore, epigraphic evidence from the walls of Constantinople suggests that in the fifteenth century wealthy individuals, such as Manuel Iagaris and Manuel Bryennios Leontaris, funded repair works on the walls. For instance, an inscription from the gate of Pege, which is dated May 1433 or 1438 reads:

> This God-protected gate of the life-giving spring (Pege) was restored with the co-operation and at the expense of Manuel Bryennios Leontaris in the reign of the most pious emperors John and Maria Palaiologina.

Another inscription dated 1448 reads: 'This tower and curtain wall were restored by George (Branković), despot of Serbia in the year 6956.'[24]

Moreover, practical considerations and the inability of the throne to support Anatolian cities under blockade by the Turks meant that their defence relied on local means and authorities. The bishops Niphon of Kyzikos and Theoleptos of Philadelphia personally directed the defence and fortification of their cities against the Turks.[25] Theoleptos, in the blockade of Philadelphia in 1310, supervised the feeding of the population and carried out negotiations with the Turcoman principality of Germiyan. According to an inscription in Kütahya, which was the capital of Germiyan, Theoleptos agreed that Philadelphia should pay tribute to Yakub Alişir if the blockade were lifted.[26] Officially, the Byzantine church did not allow the clergy to participate in military activities. However, considering the needs of the moment, the patriarch Athanasios I, in one of his letters in which he attacks the bishops who have abandoned their sees and made their residence in Constantinople (1304–1305), approves Theoleptos' actions.[27] In addition, the increased sense of insecurity led large monasteries, such as

[23] Kantakouzenos, I, 185.

[24] Van Millingen, *Byzantine Constantinople*, 106–107, 193, whose translations these are. For Iagaris and Leontaris see *PLP*, 7810, 14669.

[25] Pachymeres, IV, 425; Gregoras, I, 221; P. Schreiner, "Zur Geschichte Philadelpheias im 14. Jahrhundert (1293–1390)," *OCP* 35 (1969), 385–393.

[26] A.C. Hero, "Theoleptos of Philadelpheia (ca. 1250–1322): From Solitary to Activist," in S. Ćurčić – D. Mouriki, (eds.) *The Twilight of Byzantium* (Princeton, 1991), 35; I. Mélikoff, *Germiyān-Oghullari*, in *EI*², col. 989.

[27] Athanasios I, Patriarch, *The Correspondence of Athanasios I Patriarch of Constantinople*, A.-M. Talbot (ed.), (Washington, 1975), 56.

Chilandar and Vatopedi in Athos, to construct towers which reveal an attempt to consolidate a system of defence which could have made them independent of outside military assistance. Furthermore, the end of the late medieval period in the Balkans witnessed the widespread use of towers constructed by individuals who had prospered through commerce and trade. For instance, in fifteenth-century Constantinople a significant number of wealthy and powerful men built houses which included towers. Safety and display were the motives for the construction of these private towers and it is worth noting that the increased sense of insecurity affected the construction not only of private effigies but of palaces as well.[28]

Regardless of whether urban defences were built, funded, and maintained by imperial authorities, wealthy aristocrats, or local authorities and the church, the success of the late Byzantine system of defence depended heavily on the strength of the fortified cities, described by the sources as *phrouria* and *kastra*. Very often, the defence of major cities was supported by smaller fortresses around them, such as Gynaikokastro and Rentina, which are attested in almost every military operation conducted around Thessalonica.[29] With the exception of the western Asia Minor frontier during the Nicaean period, where the protection of this quite peaceful frontier relied on local troops, fortified cities cannot be seen as a distinct frontier defence system. The reason is that throughout the late Byzantine period and in the Palaiologan period in particular, very few, if any, of the imperial territories remained unaffected by warfare. Fourteenth and fifteenth-century Byzantium was unable to effectively control a large linear frontier zone. Consequently, Byzantium's defence relied heavily on the strength of fortified cites which were scattered throughout the empire. Therefore, a more narrowly defined notion of security was introduced.[30] For instance, in the last decades of the thirteenth century and during

[28] S. Ćurčić, "Late Medieval Fortified Palaces in the Balkans. Security and Survival," *Mnemeio kai Perivallon*, 6 (2000), 11–48; idem, *Architecture in the Balkans from Diocletian to Süleyman* (Yale, 2010), 518–527.

[29] Kantakouzenos (II, 235) states that the garrison of Gynaikokastron was under the command of the governor of Thessalonica. For Rentina and Gynaikokastron, see N.K. Moutsopoulos, *Άγνωστα Βυζαντινά κάστρα της Μακεδονίας* [Unknown Byzantine Castles of Macedonia] (Thessalonica, 2004), 194–198; idem, *Ρεντίνα* [Rentina], 4 vols. (Thessalonica/Athens, 1995–2001), passim; Tourta, A. "Fortifications of Gynaikokastron," in S. Ćurčić – E. Hadjitryphonos (eds.), *Secular Medieval Architecture in the Balkans, 1300–1500, and its Preservation* (Thessalonica, 1997), 110–113.

[30] Ćurčić, "Architecture in the Age of Insecurity," 27.

the intense warfare with Serbia, the Byzantine defence relied on the possession of important fortresses and fortified cities such as Ochrid and Veles. It seems also that the late thirteenth century witnessed an intensive construction of fortifications on the part of the Byzantines in an attempt to thwart the threat posed by Uroš II Milutin.[31]

One of the most important functions of the heavily fortified cities was to provide shelter to people during enemy raids. The aim was to move the people from the countryside to the cities, quite often bringing their flocks and supplies with them, so that the damage caused by the enemy would be minimised and the enemy would then not find supplies of food, etc., to live on. Consequently, the prolonged permanence of the enemy forces on imperial soil without the provision of adequate supplies would give sufficient time to imperial mobile troops to assemble and organise their counter-attack against an exhausted enemy. Moreover, the Byzantines knew that raiders would be much more concerned to return to their bases unscathed with their booty than risk a battle. In some cases, the Byzantines had received in advance sufficient intelligence about forthcoming raids, so that there was time to assemble the people in the fortified cities. According to Kantakouzenos, a Tatar raid in 1322 caused small damage because the heads of the Thracian cities, informed in advance, gathered the inhabitants inside the city walls.[32] He also states that, in 1323, the Bulgarians raided Thrace for twelve days and they reached Vera. However, they did not cause any serious damage to the Byzantines because special care had been taken and the farmers were protected inside the fortresses and cities.[33] Gregoras states that he encountered many difficulties on his way to Serbia as an envoy in 1326/1327, because the countryside was deserted and the people in Thrace had fled to the fortresses at rumours about an imminent Tatar raid.[34]

The surviving information indicates that it was ideal for the Byzantine armies to encounter their enemies loaded with booty or dispersed and disorganised in search of booty. The Tatar raid led by

---

[31] See the analysis of M. Popović, "Les forteresses dans les régions des conflits byzantinoserbes au XIVe siècle," in N. Oikonomides (ed.), *Byzantium and Serbia in the 14th Century* (Athens, 1996), 67–87.

[32] Kantakouzenos, I, 188.

[33] Kantakouzenos, I, 179; Gregoras, I, 430.

[34] Gregoras, I, 374–375, Nikephoros Gregoras, *Nicephori Gregorae epistulae*, 2 vols. A. Leone (ed.) (Lecce, 1982), 99–115.

Taitah and Toglu Turgan, in 1323, lasted forty days.[35] The emperor could not assemble the army because the soldiers were defending their own home cities. Therefore, for fifteen days Andronikos III followed the Tatars with whatever troops he had available and returned to Adrianople without having accomplished anything. His next step was to order Kantakouzenos to lead a force against the raiders. In Kantakouzenos' account, in the village of Proumousoulou he took by surprise a part of the Tatar army loaded with booty. In the meantime, Andronikos set out from Adrianople and assembled the troops from the individual cities and their leaders and, with Kantakouzenos, encountered the remaining Tatars at the river Tountzas.[36]

Similarly, in 1330, Andronikos III took by surprise a Turkish raiding force which was plundering in Thrace and had been divided into two groups.[37] Gregoras relates that when in 1337, the Turks of Saruhan raided Thrace, the emperor was in Didymoteichon with no significant force to confront the raiders. Instead, leading a force of 150 soldiers, he laid ambushes. Taking advantage of the fact that the Turks were dispersed in small groups, he managed to get back the booty from each group of the enemy troops that he encountered.[38] In the summer of the same year and while Andronikos III was preparing for the campaign in Epiros, Orhan reached the suburbs of Constantinople. According to Gregoras, his aim was to plunder the crops in the area around Constantinople and to seize two fortresses very close to Constantinople, to use them as a base for a future attack. Gregoras claims that Kantakouzenos was sent by Andronikos III with only seventy cavalrymen. During the night he found the Turks dispersed and disorganised. He claims that the Byzantines managed to kill 1000 of them by the next day, while the emperor leading another force seized fourteen enemy ships.[39] Kantakouzenos states that he was successful because he attacked the Turks while they were dispersed and looking for booty.[40]

In case of very serious threats, extreme measures were taken, which seem to have caused trouble and conflicts. According to a document

---

[35] G. Moravscik, *Bizantinoturcica*, 2 vol. (Berlin, 1958), II, 315.
[36] Kantakouzenos, I, 188.
[37] Kantakouzenos, I, 435.
[38] Gregoras, I, 538.
[39] Gregoras, I, 540–541.
[40] Kantakouzenos, I, 505–506.

dated sometime between 1234 and 1239, when Henry of Flanders invaded Asia Minor from Constantinople (1211), George Laskaris, the brother of Theodore I, evacuated the theme of Thrakesion and sent the population to *kastra* and to mountainous and fortified areas. It seems that the return of the people after the end of the conflict caused disputes over the possession of property. According to this document, John III asked his uncle and *doux* of the Thrakesion, John Angelos, to intervene and solve the dispute between the monastery of Roufianon and the *paroikoi* of the *lizios kavallarios* Syrgares over the possession of a farm, of which both sides claimed ownership. More specifically, the monastery claimed that the *paroikoi* of Syrgares had at the end of the evacuation seized the field illegally. The *paroikoi* supported by Syrgares claimed that they had received that land from their parents, before the evacuation.[41] It is not possible to say how many other similar problems arose when the state decided to take measures of this kind.

In other similar examples, after the battle of Apros (July 1305), when the Byzantine troops under Michael IX were crushed by the Catalans, Andronikos II had contemplated scorching the countryside around Selymvria. The reason behind this plan was that the Catalans and their Turkish allies were living off the land cultivated by the Byzantines, who had been forced to flee, while after the battle of Apros Michael IX commanded a very small force, which could not protect the local population adequately.[42] Shortly afterwards, in the spring of 1306, the council around Andronikos II decided to prohibit the cultivation of land. Although the farmers were able to go outside the walls, the decision was based on fear of the continuous attacks carried out by the Catalans, and the danger that they might profit from the crops and extend their stay in the area. The Catalans, as a company of professional mercenaries, did not cultivate the land, and as Pachymeres states, famine was a threat for both the Byzantines and the enemy.[43] This measure was accepted not without reaction, as the letters of the patriarch Athanasios I indicate.[44] Such measures were taken in extraordinary circumstances. The empire had to face the Catalans, while the cities in Asia Minor were under pressure from the Turks, and the Bulgarians had taken advantage of the general instability to carry out

41 MM, IV, 35–41.
42 Pachymeres, IV, 669.
43 Pachymeres, IV, 691; Muntaner, II, 530.
44 *The Correspondence of Athanasios I*, 158–160.

raids on Thrace. As Pachymeres states, 'our army could not oppose the multitude of our enemies.'[45]

As far as the garrisoning of fortresses is concerned, the sources mention the *tzakonike phylaxis* (tzakonian guard).[46] The Tzakonian guards are mentioned as early as the tenth century in the works of Constantine VII.[47] Although in the thirteenth century the name applied to the light-armed soldiers whom Michael VIII had transferred from the Peloponnese to Constantinople, the tzakonian service, as described by Constantine VII, continued to exist. This was the duty of watching over the fortress and its gates; it was performed not only by soldiers but also by civilians.[48] Referring to his attack on Veroia in 1350, Kantakouzenos remarks that his troops encountered 'citizen guards' (*phylakes politai*).[49] It seems also that the *pronoia* holders who performed military service, the *stratiotai*, were not expected to perform tzakonian service. This is shown by a document edited by the quasi-independent ruler of Thessaly, Michael Gavrielopoulos.[50] The sources indicate that in an emergency, more guards were urgently recruited. Gregoras states that when Andronikos II in 1307 became aware that the Catalans were planning to advance westwards in the following spring he,

> Selected and sent to Macedonia generals, who were not ignorant of warfare, to recruit in Macedonia those who would be competent guards of the Macedonian cities if the enemy decided to besiege them, to gather and assemble from the suburbs abundant food supplies and to store them in the cities, and to take care of every other matter, so that during the siege the guards would not find the internal siege of starvation more threatening than the siege from outside.[51]

One of these generals was Michael Chadrenos, who contained the Catalan attacks in Thessalonica and carried out successful counterattacks against them.[52]

---

[45] Pachymeres, IV, 663.

[46] For the social origins, recruitment and finance of the garrison troops see Bartusis, *The Late Byzantine army*, 292–302.

[47] Constantine VII Porphyrogenitus, *De Ceremoniis aulae byzantinae*, 2 vols I. Reiske (ed.), (Bonn 1829–30), I, 696.

[48] M. Bartusis, "Urban Guard Service in Late Byzantium: The Terminology and the Institution," *Macedonian Studies* 2 (1988), 54–57.

[49] Kantakouzenos, III, 307.

[50] MM, V, 260.

[51] Gregoras, I, 246.

[52] *AG*, II, 193–195, 226–227; For the activities of the Catalans in Macedonia and the Byzantine reaction see Laiou, *Constantinople and the Latins*, 220–226.

The importance of the local garrisons for the defence of the empire is shown by the fact that the defence of Thrace against the Catalans relied exclusively on the strength of the garrisons of the individual cities and on raids of Byzantine troops aiming to damage the resources of the Catalans. For instance, the troops of Tzouroulou attacked the Catalans in Raidestos, seizing considerable booty. Pachymeres provides some further information concerning the strength of the local garrisons, stating that the Catalans could not cope with the endurance of the local population. In 1306, the defence of Adrianople was made up of 150 soldiers from Asia Minor, and of the soldiers around Senachereim Angelos and a certain Choumnos. The Catalans besieged Adrianople for eight days, using siege engines and a tower with wheels, which the besieged people managed to destroy. Pachymeres relates that after this failure the Catalans besieged Pamphilon and failed again. The tower of Raidestos was defended by a force of 300, which carried out many sorties, damaging the Catalans. When the Catalans and their Turkish allies attacked Vizye, the city was defended by 200 cavalry, led by Oumbertopoulos.[53] It should be noted that the Catalans, unlike the Turks in Asia Minor, did not have the time or the patience to starve the Byzantine cities into submission, although they used siege weaponry. They were fighting in a hostile area without a stable supply of resources, living merely on the booty collected in their raids.[54]

In cases of emergency, whole cities were mobilised. In 1342, Kantakouzenos' army in Didymoteichon included 8,000 archers in the lower city, under the command of eight heads.[55] Unless Kantakouzenos provides us with a grossly exaggerated figure, the magnitude of this number and the ensuing conflict between the people of the lower city and the soldiers in the upper city leads to the conclusion that most probably those archers were inhabitants of Didymoteichon capable of bearing arms and possibly, as Bartusis assumes, reinforced by various irregulars from the surrounding area.[56] Kananos, in his account of the siege of Constantinople in 1422 by the Ottomans, states that not only the soldiers and those who know the art of war, but also the *archontes* and all the common people and the clergymen, as well as many

[53] Pachymeres, IV, 665–667, 685, 689, 693.
[54] Muntaner, II, 463.
[55] Kantakouzenos, II, 195.
[56] Bartusis, *The Late Byzantine Army*, 297.

women, provided the men with stones, water and torches and help for the injured.[57]

Those in charge of the fortified cities were in most cases military commanders, who exerted both military and civil authority. The sources indicate that the local head (*kephale*) supervised the defence in case of an enemy raid and was assisted by the *tzaousios* and the *kastrophylax* (castle guard).[58] This strategy required military commanders able to assess the strength and quality of their troops, as well as the nature and strength of the enemy. Moreover, they had to be able to co-ordinate the local population and their troops effectively, imposing their will on the civilian population, who might urge the head to carry out attacks against the raiding enemy, overestimating the strength of their own side. It seems that to impose the proper co-ordination between the local population and the army and to resist the pressure of the local population was not an easy task for the Byzantine commanders. For instance, the general of Mesemvria, the *kouropalates* Oumbertopoulos, is praised for defeating, on his own initiative, a Tatar raiding force in 1284.[59]

However, in most cases, the local heads were either ordered by the emperors not to attack the invading enemy, or are criticised for carrying out an attack against the enemy and being defeated without taking the proper precautions and without assessing the strength exerted against them. It is noticeable that in such cases the sources indicate that, although the local heads and their troops did not wish to confront the enemy on the battlefield, they did not manage to impose their will on the local population. The *megas tzaousios* Oumbertopoulos (he should not be confused with the above *kouropalates*), as head of Vizye, in 1307, with a force of more than 200 cavalrymen, had intended to remain inside the walls and not to challenge the attacking Catalans, led by Ferran Ximenez, and their Turkish allies. According to Pachymeres, Oumbertopoulos' forces were outnumbered. However, the local population were willing to fight and a large force was assembled. The people told Oumbertopoulos that if he consented they would be willing to fight. Pachymeres criticises Oumbertopoulos, saying that he grew overconfident and, lured by the large number of fighters assembled, he decided to attack. The description of the ensuing battle makes clear

---

[57] Kananos, 19–20.

[58] For the duties of the *tzaousios* and the *kastrophylax* see Bartusis, "Urban Guard," 57–60.

[59] Pachymeres, III, 93.

that the lack of discipline, coherence and professionalism was the main cause for the Byzantine defeat.[60]

Another example of the effect of pressure from the local population on the local commanders during an enemy incursion is provided by Kantakouzenos. In 1342, a Tatar force remained inactive because it found nothing to plunder, due to the destruction caused by the civil war, and because everyone had withdrawn from the countryside and found refuge in the cities. The Tatars set up camp outside the town of Skopelos in Thrace. Kantakouzenos criticises the people of Skopelos for forcing the local governor, a certain Michael, to attack. Michael initially refused to make the first move, but the pressure of the locals seems to have been too great for him. Kantakouzenos, using a phrase characteristic of his social views as a member of the military aristocracy, states, 'the soldiers decided to join the people in the attack against the Mongols, choosing to be killed fighting the barbarians and not to be killed by the mob.' He is also very critical of the fact that this army, completely lacking in discipline, made the fatal mistake of attacking the Tatar cavalry on flat terrain. Eventually, only the Byzantine cavalry escaped the massacre because it managed to run away.[61]

Siege warfare: The evidence of the written sources

The available source material indicates that sieges were the dominant form of warfare in the period under discussion. However, very little is known about the siege equipment and techniques employed by the Byzantine army and its enemies, since the sources of the period do not provide accurate descriptions of siege weapons and use inconsistent and archaic terminology. The problem of terminology is reflected in Kritoboulos' description of cannon in the fifteenth century. After mentioning that cannon was a new weapon which old emperors neither possessed nor knew anything about, he points out that no ancient name has been found for this engine, unless someone calls it *helepolis* (city-taker) or *apheterion* (thrower). Its common current name, as he concludes, is *skeue* (apparatus). In his account of the siege of

[60] Pachymeres, IV, 693.
[61] Kantakouzenos, II, 303.

Constantinople in 1422 by Murad II, Kananos uses the terms *skeue* and *voumbarda* to describe the Ottoman canons.[62]

Akropolites, who accompanied John III and Theodore II on their campaigns, relates that both emperors used siege engines. He also comments that John III preferred to expand the Nicaean state through successful sieges and blockades rather than through pitched battles. In his words, John III,

> Was patient in battles and was not happy with pitched battles, fearing the fickleness of war and taking into account the uncertainty of these things. By being patient and spending the spring in the enemy land and carrying over into the summer, and even into the autumn, and sometimes passing the winter, he gained victory, since the enemies were exhausted due to the pressure and patience of the Emperor.[63]

It is possible to find even more specific examples in the *History* of Akropolites. In 1223/1224, after defeating the army of the Latin empire of Constantinople in Poimamenon John III captured the fortresses of Poimamenon, Lentiana, Charioros and Ververiakon.[64] According to Akropolites, these operations involved long sieges which were conducted in the middle of winter and the Nicaeans used *helepoleis* (city-takers). Eventually, some of these towns surrendered to John III because their supplies were exhausted and they did not expect any military aid, while others were captured by force. Describing John III's siege of Thessalonica in 1242, Akropolites states that the large size of the city made it impossible to put in place *helepoleis* and other siege engines and conquer it by this kind of warfare. Therefore, the emperor plundered all the surrounding area. While the *helepoleis* proved ineffective in Thessalonica, in 1256 the use of these, together with other unidentified war machines, facilitated the capture of the fortified towns of Peristitza, Stenimachos and Krytzimos, in the Rhodope mountains, which immediately after John III's death were seized by the Bulgarians. In 1259, shortly after Michael Palaiologos' promotion to despot, Akropolites remarks that John Palaiologos was determined to conquer the city of Deavolis. For this reason, he set up *helepoleis*

---

[62] Kritoboulos, *Critobuli imbriotai historiae*, P. Reinsch (ed.), (Berlin, 1983), 45–46; Kananos, 6–7.

[63] Akropolites, I, 104.

[64] For these cities which are in the area around Lake Aphnitis see W.M. Ramsay, *The Historical Geography of Asia Minor* (London, 1890 repr. New York, 1972), 158.

and all kinds of engines of war. Eventually, after many people died inside the town, its authorities surrendered it to the Byzantines.[65]

The *History* of Akropolites indicates that sieges were a significant part of the military operations undertaken by the Nicaean armies. The importance of successful sieges cannot be overestimated. The victory of John III over the Latins at the battle of Poimamenon would have brought limited results unless it had been followed by successful sieges of the towns which were under enemy control. In 1255, the victory of the Nicaean army against the Bulgarians on the battlefield facilitated the capture of Veroe (modern Stara Zagora) by Theodore II.[66]

As far as the siege techniques of the Nicaeans are concerned, Akropolites remarks that the armies of John III and Theodore II used *helepoleis* and other war machines. What were these *helepoleis*? It has been argued that, since the sixth century Byzantine authors had used the term *helepolis* together with generic references to 'stone-throwing machines' to refer to traction trebuchet.[67] Moreover, it has been maintained that, before the Laskarids, the reigns of John II Komnenos (1118–1143) and Manuel I Komnenos saw a substantial increase in Byzantine reliance on siege warfare and on the use of counterweight trebuchet, which the sources call *helepolis*.[68] Twelfth-century authors such as Niketas Choniates, John Kinnamos and John Tzetzes describe the effectiveness of the trebuchets in breaching fortifications and terrifying the inhabitants of cities besieged by the Byzantines.[69] That Akropolites distinguishes the *helepoleis* from the rest of the war machines implies their important role in the sieges led by the Nicaean rulers. However, the absence of an elaborate description prevents us from reaching safe conclusions regarding their form and impact. It is not safe to conclude that Akropolites follows the twelfth-century historians and uses the term *helepoleis* to refer to trebuchets. One

[65] Akropolites, I, 36, 66, 113, 167. For these fortresses in Rhodope see Asdracha, *Rhodope*, 166–170.

[66] Akropolites, I, 111–112.

[67] G.T. Dennis, "Byzantine Heavy Artillery: The Helepolis," *GRBS* 39 (1998), 104; P.E. Chevedden, "The Invention of the Counterweight Trebuchet: A Study in Cultural Diffusion," *DOP* 54 (2000), 75.

[68] Dennis, "Heavy Artillery," 110; Chevedden, "The Invention of Counterweight Trebuchet," 76; J. Birkenmeier, *The Development of the Komnenian Army* (Leiden, 2002), 182–183. For the origins of the counterweight trebuchet see K. De Vries, *Medieval Military Technology* (New York, 1992), 137–138.

[69] See Dennis, "Heavy Artillery," 110–114; Chevedden, "The Invention of Counterweight Trebuchet," 76–78, 80–81, 86–87, 92–93, 104.

wonders whether the *helepoleis* mentioned by Akropolites are identical to the *pierieriès et mangonniaus* (stone-throwers and mangonels) which, according to Villehardouin, were used by the Bulgarian armies of Kalojan (1197–1207).[70]

Moreover, it is interesting that while Akropolites states that due to the great size of the city John III did not use *helepoleis* against Thessalonica, in his account of the sack of Thessalonica in 1185 by the Normans, its bishop, Eustathios, tells us that the besiegers on the eastern side of the city used many small stone-throwing machines (*petrovolai*) which breached its walls. One of the largest ones was named 'daughter of the earthquake' because of its effect and noise. The Normans on the western side of Thessalonica used newly invented *helepoleis* which were too large for them to handle properly and had no effect on the defence of the city.[71] Does this mean that the Nicaean *helepoleis* were of a different nature and less effective than the *helepoleis* deployed by the Normans in 1185 and that the Normans employed better engineers? Although the Norman kingdom of Sicily reached a high level of expertise in using siege equipment in the twelfth century and its leaders were proficient in siege warfare, the lack of sufficient information prohibits safe conclusions.[72]

Michael VIII's unsuccessful siege of Galata in April 1260 is the only siege operation described in detail by Pachymeres. Pachymeres points out the strategic importance of the operation by commenting that in holding the fortress one held Constantinople. He records that the emperor pitched his tent on a hill to watch the whole operation and to be seen by the enemy in order to frighten them. One of the first actions of the Byzantines was to set up 'wall-fighting engines' (*teichomachika*). Among the Byzantine soldiers Pachymeres distinguishes the archers from the area around Nicaea, who, as he comments, by shooting accurately did not allow any of those inside the walls to lean out. Pachymeres implies that an important disadvantage of the army of Michael VIII was that it was not supported by a fleet which could prevent the besieged people from receiving reinforcements. As a result, every day more 'Italians' entered Galata through its sea gate.

---

[70] Villehardouin, 169–170, 178, 185.

[71] Eustathios of Thessalonica, *La espugnazione di Tessalonica*, S. Kyriakidis (ed.), (Palermo, 1961), 72; Chevedden, "The Invention of the Counterweight Trebuchet," 94–95; R. Rogers, *Latin Siege Warfare in the Twelfth Century* (Oxford, 1992), 122.

[72] See Rogers, *Latin Siege*, 122–123.

Pachymeres comments that the Latins defended the walls vigorously and that,

> Many of the men outside the walls fell, hit from unseen places because standing on their feet the new troops appeared in turns for a while, drawing with ease the arrow shooting apparatus (*iovola skeue*), with which they were familiar.

The Byzantines used stone-throwers (*petrovola*). The besiegers defended the city against the stone-throwers by collecting vine branches and covering the damaged parts of the walls with them. As for themselves, they headed to the covered galleries and protected their heads with vine branches. Pachymeres continues his account of the siege by commenting on the motivation of the two opposing armies. He remarks that the Nicaeans were encouraged by the large size of their army, their strength and experience and the presence of the emperor. The defenders were encouraged by their high morale and high intelligence, as well as by the fact that they were incited by an audacity born out of fear due to danger they faced. According to Pachymeres as the siege dragged on, it led to a feeling of shame in those around the emperor, that such a large army was unable to capture the fortress. But at the same time, the Italians wished to appear to be victorious over the emperor, although they were outnumbered. Pachymeres concludes that the large number of casualties caused by the arrow shooting machines of the besiegers (*iovola mangana*) and the rumour that a fresh and powerful army was coming to reinforce the defenders led Michael VIII to lift the siege.[73]

Pachymeres was, however, not an eye-witness and his sources are unknown. Nonetheless, even if we doubt the accuracy of his account, what he describes seems to be representative of the siege operations of the period under discussion. However, by consistently using archaic terminology Pachymeres makes it difficult to be sure what weaponry was used by the besiegers and the besieged. The siege engines of the Byzantines were supported by Nicaean archers, who according to Pachymeres were very effective, while the besieged used *iovola skeue*, which must have been crossbows. The *iovola mangana* which, according to Pachymeres, killed many of the besiegers are likely to have been ballistae. According to Chevveden, the terms *manganon* and *manganinkon* were used by some authors to denote the pole-framed

[73] Pachymeres, I, 171–175.

trebuchet and by others every type of trebuchet.[74] In his investigation of the military treatise compiled by the general Nikephoros Ouranos in ca 1000, McGeer concludes that *manganika* were trebuchets which were used not for the breaching of walls but mainly to remove the defenders from the ramparts and complemented the units of archers and slingers.[75] However, the *mangana* mentioned by Pachymeres were used by the besieged. It seems that the use of vine branches to absorb the shock of missiles was not uncommon. In the siege of Constantinople in 1204 the Venetians used vine to protect their ships from Byzantine artillery.[76] Moreover, the *Taktika* of Nikephoros Ouranos suggest the construction of hut-like shelters (*laisai*) which would protect the attackers from the missiles of the besieged. They were probably made of vines and branches.[77]

In his account of the expansion of the Turcoman principalities in Asia Minor, Pachymeres indicates that the Turkish armies relied on successful blockades of Byzantine cities and on the inability of the throne to send relief armies to reinforce the besieged. Consequently, in spite of being unable to use sophisticated siege techniques, the armies of the Turcoman principalities were not caught between garrisons and enemy field armies. Therefore, they were able to starve the Byzantine cities out. This shows that the possession of heavily fortified cities alone could prevail against enemy raids which lasted for no more than a few days or weeks and aimed at the acquisition of booty. However, it was an ineffective method of defence against enemies whose aim was the permanent capture and possession of Byzantine territories and cities. Pachymeres comments that initially the Turks raided and plundered the Byzantine possessions. They destroyed the economic basis of Byzantine cities and gradually took control of the surrounding countryside. As a result the people in the cities were impoverished and forced either to submit or to flee. Consequently, it was easy for the Turks to return and occupy the land deserted by their opponents.[78] Pachymeres indicates that by the first decade of the fourteenth century the communication system in Asia Minor was in a state of collapse.

---

[74] Chevedden, "The Invention of Counterweight Trebuchet," 79.

[75] McGeer, *Sowing the Dragon's Death*, 160; idem, "Byzantine Siege Warfare in Theory and Practice," in I.A.Corfis – M. Wolfe (eds.), *The Medieval City under Siege* (Woodbridge, 1995), 128.

[76] Robert de Clari, 73.

[77] McGeer, *Sowing the Dragon's Death*, 158.

[78] Pachymeres, I, 293.

He relates that it was impossible to go on foot from Constantinople to Herakleia in Pontos. In another example, he remarks that there was only one open road to Nicaea, which was full of dangers, while Nikomedia was ready to surrender.

Referring to events that occurred three decades later, Kantakouzenos complements Pachymeres' account of the Turkish methods of conquest. In 1331, during the preparations for a campaign against the Bulgarians, Andronikos III received news that the Ottomans were ready to attack Nikomedia. Kantakouzenos relates that Nikomedia could not be taken in battle due to the strength of its fortifications and its geographical position. The only way it could be taken by the enemy was if it ran out of supplies. Kantakouzenos adds that the Turks were well aware of this and therefore they were planning to seize the road by which the city received its food supplies. The emperor was forced to put off the preparations to march on the Bulgarians and went to relieve Nikomedia. Gregoras adds that in 1337 Nikomedia surrendered to the Ottomans after being starved into submission.[79] The disruption of the lines of communication, mainly the result of the Turkish control over the countryside, made it very difficult for the Byzantine armies to march and reinforce the cities of Asia Minor.

Pachymeres' account of the Turkish raids and the submission of Asia Minor indicates that the Turks followed a well-established strategy. It was common for medieval armies to attempt to take an enemy city by pillaging its countryside and destroying its economic basis. Deprived of access to their supplies, the Byzantine cities would be demoralised and starved. Pachymeres indicates that it was not uncommon for Byzantine cities which were blockaded by the Turks to try to work with the besiegers and his account provides examples which reflect the complex relationships that developed between the blockaded Byzantines and their besiegers. In 1303 the people of Sardis, pressed by the lack of water accepted a proposal made by the head of the Turcoman principality of Menteşe whose forces were besieging Sardis. The Turks were anticipating a Mongol attack and were willing to compromise. As a result of this agreement the Byzantines were allowed to move beyond the walls to take care of their own affairs, 'while the Turks promised they would not harass them but they would

[79] Pachymeres, I, 285, II, 403–405, IV, 453, 463, 475–477; Kantakouzenos, I, 459; Gregoras, I, 545. For the fortifications of Nikomedia see Foss, *Nikomedia*, 29–43.

continue exploiting others by attacking them in the fashion of brigands.' Furthermore, Sardis accepted a great number of Turks who occupied the citadel of the city. This was not actual co-existence as long as the Byzantines and the Turks were separated by the walls of the citadel and used a single gate to carry out commercial transactions. When the possibility of a Mongol attack faded, both sides made plans to resume the attack on each other. Finally, Pachymeres states that the Byzantines,

> Judged that it was more profitable to act against the Turks rather than to sustain an attack, and called the head of the Roman troops in the area, Nostongos Doukas, who took the Turks by surprise and defeated them.[80]

According to Pachymeres, due to the long siege Tripolis was in desperate need of supplies.[81] However, since communications had been disrupted, there was no way to transfer any supplies or reinforcements. Therefore, the Byzantines decided to reach an agreement with the Turks, in order to be able to receive what they needed. Pachymeres states that,

> As is customary in commerce not only they (the people of Tripolis) were able to come out of the city to buy their necessities, but they also allowed the Turks to get in freely to sell their products. Because this happened so many times it gave the Turks the idea to seize the city. They contacted traitors, giving them the necessary information and fixed the day of the attack.[82]

The information provided by Pachymeres and Kantakouzenos is confirmed by the late fifteenth-century chronicle compiled by a soldier of Murad II, Aşık Paşazade, who claims that his source of the early history of the Ottomans was a chronicle written by Yahşı Fakıh, a

---

[80] Pachymeres, IV, 443. For the dating of these events see Failler, "Les Émirs Turcs à la conquête de l' Anatolie au début du 14e siècle *REB* 52 (1994), 81. For Late Byzantine Sardis see C. Foss, *Byzantine and Turkish Sardis* (Harvard, 1979), 81–89.

[81] For the fortifications of Tripolis see Foss-Winfield, *Byzantine Fortifications*, 299–302.

[82] Pachymeres, IV, 478. This atmosphere of co-operation and conflict between the Byzantines and the Ottomans is also reflected in the Ottoman sources. See K. Hopwood, "Low-level diplomacy between Byzantines and Ottoman Turks: the Case of Bithynia," in J. Sheppard - S. Franklin (eds.), *Byzantine Diplomacy* (London, 1994), 51–55. For a comparison between Pachymeres and Turkish sources see A. Failler, "Les Émirs," 104–112.

son of an imam of Orhan.[83] This Ottoman chronicle confirms that the Ottoman expansion in Bithynia relied on long blockades and on the destruction of the countryside around Byzantine cities. To capture Prusa, the Ottomans built two forts to blockade it. Similarly, they took two forts very close to Nicaea in order to blockade the city and destroy its countryside. Moreover, Aşık Paşazade relates that in 1326 a Byzantine administrator in Prusa told Orhan that the city surrendered because the Ottomans had already captured the villages around it.[84]

Until the middle of the fourteenth century the Ottoman and non-Ottoman Turks relied on close blockades which compelled the Byzantine cities to seek a negotiated surrender. Nonetheless, this does not mean that fighting never occurred on and around the walls of Byzantine cities. In his account of the capture of Tralleis by the army of the emirate of Menteşe, Pachymeres points out that the exhaustion of the water supplies combined with the fact that the Turks had managed to undermine its walls led to the surrender of the city.[85] When in 1307, the Ottomans tried to storm Trikokia, a fort very close to Nicaea, they were halted by the accuracy of the shots of the Byzantine archers who were posted on the walls. Eventually, the Ottomans succeeded in filling the ditches in front of the walls with sand, wood, rocks and trees. They reached the walls and captured them, killing many defenders.[86]

The *History* of John Kantakouzenos shows that sieges were the dominant form of warfare during the civil wars of the first half of the fourteenth century and during the reign of Andronikos III as sole emperor. Kantakouzenos gives the impression that the civil war of 1321–1328 was characterised by the competition of the rival parties over the support of city governors and of the local aristocracy. More often than not, this involved negotiations and not sieges. The siege of Apros led by Andronikos III in 1322 is one of the few operations of this kind described by Kantakouzenos. If we believe the figures he provides, Apros was defended by 100 cavalry, many archers and slingers. This force was supplied by 220 cavalry, 200 archers and 300 crossbow-

[83] H. İnalcık, "How to Read Ashik Pasha-Zade's History," in C. Imber – C. Heywood (eds.), *Studies in Ottoman History in Honour of Professor V.L. Mènage* (Istanbul, 1994), 141–144.

[84] Aşık Paşazade, *Osmanoğulları' nın tarihi*, K. Yavuz – M.A.Y. Saraç (eds.), (Istanbul, 2003), ch. 18, 22, 23, 32.

[85] Pachymeres, II, 597–599.

[86] Pachymeres, IV, 703.

men from Constantinople. Kantakouzenos adds that, apart from these troops, many people turned up from the surrounding area who wished to take advantage of the war situation. Apparently, he is referring to looters. He also comments that Apros' garrison was a large force.[87] Andronikos III tried to attract the defenders to his side by promising privileges. When the negotiations failed and, as Kantakouzenos writes, the besieged started launching arrows and stones from the walls, the emperor ordered his army to construct ladders and storm the town. Kantakouzenos remarks that although the fighting lasted 'until the seventh hour of the day' nobody died. It was only after the end of the battle that a Latin mercenary killed a prisoner who had injured him during the fighting.[88]

In 1323, in the interlude of peace during the civil war between the two Andronikoi, Kantakouzenos accompanied Andronikos III on the siege of Philippoupolis, which had been recently captured by the Bulgarians. Kantakouzenos comments that Philippoupolis was a large and populous city which because of its geographical position was self sufficient in peace and war. It was defended by 1,000 Alan and Bulgarian cavalrymen and by 2,000 infantrymen (*peltastas*). The siege lasted four months, which he considers a rather long duration for military operations of the kind. Kantakouzenos remarks that there was much fighting every day on the walls. Eventually, a German mercenary, who according to Kantakouzenos had been trained in the construction of siege engines, supervised the construction of a wheeled siege tower. Above the wheels there was a wooden hut (*oikia*) which could hold 100 men to push the wheels. On the top of the hut a five storey tower was built, and on each storey were eight crossbowmen. The soldiers inside the structure were hidden from the defenders. The tower became the centrepiece of a Byzantine assault and Kantakouzenos relates that when the tower reached the wall the besieged found it impossible to defend against the shots of the crossbows and abandoned their posts. However, the tower fell down by accident, a few moments before it could be attached to the walls. This incident demoralised the Byzantine army and Andronikos III lifted the siege.[89] The detailed description of the siege tower, as provided by Kantakouzenos, suggests that the use of

[87] Kantakouzenos, I, 123, I, 140.
[88] Kantakouzenos, I, 140.
[89] Kantakouzenos, I, 174.

such devices was rather uncommon in fourteenth-century Byzantium. Indeed, this is the only reference to the use of siege towers made by Kantakouzenos. In addition, its collapse indicates that the ground may not have favoured the use of a mobile tower. Moreover, it is likely that the Byzantines, perhaps due to lack of experience, failed to clean out the ditches that obstructed the route to the walls and level the ground.

The most important campaigns undertaken by Andronikos III after his accession as sole emperor involved sieges almost exclusively. Describing Andronikos III's siege of Chios in 1329, Kantakouzenos remarks that the defenders under their leader Martino Zaccaria had placed on the walls 800 heavily armoured infantrymen (*phraksamenoi*). Although both armies were deployed and ready to fight, no battle took place, since Benedetto Zaccaria joined the Byzantines, putting his brother in a difficult position.[90] In 1335, the lord of Phokaia, Domenico Catanea, captured Mytilene and the rest of Lesvos apart from Eressos and Mythimna. Andronikos III sent Alexios Philanthropenos to besiege Mytilene while he rushed to Phokaia. Without providing details, Kantakouzenos relates that except for Mytilene all the other fortresses (*frouria*) of Lesvos submitted peacefully to the imperial authority. The siege of Phokaia lasted five months and the Byzantine army received its supplies from the emirate of Saruhan. Kantakouzenos remarks that many wall battles (*teichomachiai*) were fought and that the Byzantines used a *helepolis* 'and other engines which were invented for the capture of cities.' Since in his account of the siege of Philippoupolis in 1323 Kantakouzenos refers to the siege tower constructed under the supervision of the German soldier as *pyrgos*, it is probable that the *helepolis* for Kantakouzenos was an artillery weapon. The use of siege engines proved ineffective because of the strength of the walls and the strong resistance of the defenders. Fearing that they would run out of supplies the besieged people expelled the Byzantine population of the city and transported the remaining supplies to the citadel, while the Byzantine forces were reinforced by the fleets of Saruhan and Aydın.[91]

Wishing to emphasise his personal influence on the successful siege of Phokaia, Kantakouzenos includes in his account a dialogue he had with one of the leading members of the Latin defenders, Juan

[90] Kantakouzenos, I, 375–378.
[91] Kantakouzenos, I, 476–481.

de Spinia. In this dialogue Kantakouzenos specifies that the Byzantines used a stone throwing engine and sums up the reasons why Phokaia should surrender. Mytilene, which provided the soldiers of Phokaia with their salaries, was blockaded by the Byzantines, while Genoa was unable to send reinforcements. Kantakouzenos goes on to point out that the countryside, from which the city's income came, was in the hands of the emperor, while the people of Phokaia were unable to carry out any commercial activities, since agricultural production had been destroyed by the Byzantines who had also taken over control of the sea routes.[92] Therefore, Kantakouzenos implies that, for a siege to be successful, the deployment of siege engines and fighting for control of the city walls was not enough. It was also essential to cut off the besiegers from their supply base, to ravage the enemy countryside and disrupt economic life. Kantakouzenos repeats the prescriptions of military manuals of the earlier epochs, which suggest that a siege operation should be preceded by the strangulation of the food and water supplies of the besieged town, the destruction of its countryside and harvest, and the interdiction of commerce. This does not necessarily show that Kantakouzenos was following the instructions of earlier manuals. As Dennis Sullivan states in his study of tenth-century Byzantine siege warfare, 'the manipulation of basic necessities was timeless.'[93]

It seems that the incorporation of Thessaly in 1334 and the campaign for the imposition of imperial authority on Epiros in 1339/1340 were achieved through successful sieges and blockades and it is doubtful whether any pitched battles were fought. Immediately after the death of the semi-independent despot of Thessaly, Stephen Gavrielopoulos, the Byzantine general of Thessalonica, Senacherem Monomachos marched on Thessaly and captured the fortresses previously under the control of Gavrielopoulos. Most probably he did not encounter any serious resistance. Shortly afterwards the emperor arrived in Thessaly to finish off the job and to consolidate his authority.[94] Kantakouzenos relates that the army which campaigned against the Epirots was divided into three parts in order to be able to besiege Arta, Thomokastro and

---

[92] Kantakouzenos, I, 484–490.

[93] D. Sullivan, "Tenth-Century Byzantine Offensive Siege Warfare: Instructional Prescriptions and Historical Practice," in N. Oikonomides (ed.) *Byzantium at War* (Athens, 1997), 180, with an analysis of the information provided by the military manuals.

[94] Kantakouzenos, I, 473–474.

Rogoi simultaneously. He remarks that Arta and Rogoi had to rely on the supplies they had collected the previous winter, while the siege of Thomokastro was a difficult and risky undertaking, for the besieging army was not supported by naval forces. Therefore, the townspeople were supplied by sea. Kantakouzenos' description of the sieges of Arta and Rogoi is similar to that of Phokaia. He comments that in the siege of Arta the Byzantines used every kind of war machine and many wall battles were fought on the walls. However, the strength of the fortifications and the stubborn defence put up by the besieged made the siege engines ineffective. The siege lasted from early spring until late summer.[95] According to Kantakouzenos, it was his personal intervention that persuaded the defenders of Rogoi and Arta to surrender. As was the case with the military operations in Phokaia, Kantakouzenos exaggerates his role in the re-imposition of imperial rule on Epiros and hence his account should be treated cautiously. Nonetheless, Kantakouzenos hints that the imperial army preferred to compel the Epirot cities to surrender through close blockades. He relates that in a discussion with the governor of Arta he encouraged him to surrender so as to spare the countryside from where the local aristocrats received significant incomes and the common people their food supplies. Refusing to surrender, as Kantakouzenos comments, deprived the wealthy of their revenues and threatens the common people with starvation. He adds that by not submitting to the emperor the people of Arta are destroying both themselves and their neighbouring cities, since the Byzantine army, being in a hostile territory, is unscrupulously ravaging the land.[96]

As was the case with the civil war of 1321–1328, the civil war fought in the 1340s between Kantakouzenos and the regency of John V was characterised by sieges aiming at the capture of important cities which could provide the rival parties with money, supplies and manpower. Kantakouzenos' strategy was to force Byzantine cities into submission by destroying their agricultural production. For instance, he states that after failing to persuade the city of Serres to join his side, his army ravaged its countryside 'as if it was enemy land.'[97] In 1344, his army raided the area around of Constantinople for eight days, forcing a number of cities to join his side. Shortly afterwards, he plundered

[95] Kantakouzenos, I, 511–512.
[96] Kantakouzenos, I, 519.
[97] Kantakouzenos, II, 292.

the environs of cities between Vizye and the Black Sea that had not followed his party.[98] Kantakouzenos concludes that this was the most effective method of forcing cities to follow him. He comments that in 1344, the cities that were loyal to the regency in Constantinople and were besieged by his troops had started to vacillate since they could no longer withstand the raids of his army and did not expect any military aid from Constantinople.[99] Raids were also conducted by Kantakouzenos' enemies. Being unable to lay siege against the strong fortifications of Didymoteichon, the armies of the regency of Constantinople repeatedly raided the area around the city, hoping to starve its inhabitants out.[100]

The limited resources of the rival armies meant that a long and well-organised siege was a rather expensive operation. Trickery and negotiations were preferred. When some of the guards of the citadel of Pamphilon were bribed and emptied the water containers of the garrison, the town deserted Kantakouzenos. After a short siege the citadel was captured by troops loyal to the regency of John V.[101] Finding it impossible to besiege the city of Melnik, due to its natural defences, Kantakouzenos attempted to capture it by conspiring with inhabitants who were friendly to him.[102] The only sieges of the civil war of 1341–1347 for which Kantakouzenos provides some circumstantial details are those of Thessalonica in 1342 and Peritheorion in 1343. He informs us that in Thessalonica he used a wheeled tortoise (a wooden hut) to mine the walls, while the siege of Peritheorion lasted twenty-six days and he used ladders (*klimakas*) and siege engines of every type, without, however, any success.[103]

Although he says little about the siege operations during the civil war of 1341–1347 Kantakouzenos provides more detailed descriptions of the sieges that he conducted as a sole emperor. In1350, he led a short campaign in Macedonia to recover cities that had been captured by Stefan Dušan. When Kantakouzenos' troops approached Veroia, he decided to capture the city by trickery (*klope*). Kantakouzenos considered his army too small to conduct a prolonged siege against Veroia, which, as he writes, was defended by a force of 1500 soldiers.

[98] Kantakouzenos, II, 483–484.
[99] Kantakouzenos, II, 492.
[100] Kantakouzenos, II, 302, 349. For the fortifications of Didymoteichon see Ousterhout – Bakirtzis, *Byzantine Monuments*, 94–98.
[101] Kantakouzenos, II, 188.
[102] Kantakouzenos, II, 228.
[103] Kantakouzenos, II, 197, 214, 217.

In addition, the Serbians had significantly reinforced its defences by building new walls and towers and repairing old ones. Hoping to take the city by surprising its defenders, Kantakouzenos' army carried out a night attack. The attackers used four ladders, which were hastily constructed by using ropes from the imperial tent. The initial plan to use wooden ladders was abandoned. Kantakouzenos had ordered the sailors of the flotilla following his army to construct wooden ladders, but the badly co-ordinated scheme failed.[104] After capturing Veroia, Kantakouzenos marched on Edessa. He relates that the soldiers reached the walls with ladders and makes no reference to the use of any other siege weaponry. He comments that the Byzantines captured the city after a prolonged fight on its walls which lasted from the morning to the afternoon. The capture of Edessa led the minor forts around it to submit to the emperor. It is interesting that while Kantakouzenos states that during the fighting many of the defenders of Edessa were wounded and some of them were killed, he concludes his account of the siege by stating that nobody died.[105] The low casualty figures of this battle, which are similar to those stated by Kantakouzenos with regard to the siege of Apros, show the small size of the armies involved in the fighting. After capturing Edessa Kantakouzenos marched on the town of Servia. He relates that Servia was too well fortified to be assaulted and that it was defended by Preljub, one of Stefan Dušan's leading generals, with 500 soldiers. The siege of Servia is the only operation in which Kantakouzenos mentions that the Byzantines mined the walls while the archers were shooting at the defenders on the walls.[106]

The *History* of Kantakouzenos indicates that in the first half of the fourteenth century sieges substantially outnumbered battles. Most of the battles that Kantakouzenos describes were not decisive and some of them were accidental. In 1328, 300 cavalrymen were sent from Serres to reinforce the garrison of the citadel of Thessalonica, which, however, had recently been captured by the army of Andronikos III. As a result a battle was fought and the soldiers from Serres were defeated.[107] It seems also that the relatively large-scale battles that Andronikos III fought were imposed upon the Byzantines. In Pelekanos-Philokrene

[104] Kantakouzenos, III, 123–125.
[105] Kantakouzenos, III, 128–129.
[106] Kantakouzenos, III, 132.
[107] Kantakouzenos, I, 270.

(1329), the Byzantines decided to withdraw after being unable to force the Ottomans to fight at close quarters. However, misinformation and the lack of order in the Byzantine camp enabled the Ottomans to carry out a counterattack which led to the Byzantine defeat.[108] If we believe Kantakouzenos, on the eve of the battle of Rosokastro (1332), Andronikos III, who was leading a raid against Bulgarian possessions, had reached an agreement with the Bulgarian emperor John Alexander and was ready to withdraw. However, the Bulgarians did not keep their promises and attacked the Byzantines.[109]

Kantakouzenos shows that the possession of towns and fortresses was more important and less risky than a victory on the battlefield. Cities and fortresses could supply armies with money, food, fodder and often manpower, since local soldiers could reinforce the armies of the new masters of their city. Although he participated in most of the sieges he mentions and describes, Kantakouzenos provides little information about their nature. He was more interested in emphasising his personal influence on Byzantine military successes than in providing information about the siege engines and technology employed by the Byzantines and their enemies. Nonetheless, his numerous references to ladders suggest that this was the most common siege weapon used by the armies he and Andronikos III led. The predominant use of ladders corresponds to the small scale of the siege operations described by Kantakouzenos. They could be constructed by materials close to the beleaguered area and their construction did not require advanced technological knowledge and the presence of specialised engineers. Unlike siege towers and artillery weapons, they did not take much time to prepare and since they were constructed *in situ* they did not delay the march of the army. It is impossible to know the quantity and type of the engines of every kind to which Kantakouzenos refers so often. Most probably, the repeated use of this is a literary *topos* aimed at emphasisng the effort of the Byzantine army to vanquish the enemy and should not be taken literally. It seems also that more often than not siege engines were ineffective. The sieges of Philippoupolis and Peritheorion failed, while most probably Phokaia and the cities of Epiros were taken in consequence of negotiations and blockades. The breaching of walls which could lead to the submission of the city

[108] Kantakouzenos, I, 342–363.
[109] Kantakouzenos, I, 464.

required the use of a large number of engines and consequently the mustering of a considerable military force and the recruitment of specialised men who possessed the knowledge and skills needed to operate sophisticated siege engines.[110] However, the sieges described by Kantakouzenos were not particularly violent clashes and involved small numbers of troops, which often did not exceed a few hundred men.

While the archaic terminology of Byzantine authors makes it impossible to establish a consensus about the type and impact of the siege engines used by the Byzantine armies, the anonymous author of the *Chronicle of the Morea* gives the impression that the use of trebuchets (*trimpoutseta*) was rather common in the siege warfare in the thirteenth and early fourteenth-century Peloponnese.[111] It is worth noting that, according to the *Chronicle of the Morea*, the Crusaders used three trebuchets against Monemvasia. However, this is unlikely, due to the topographical setting of the area.[112] The numerous references to the use of trebuchets made by the anonymous author show the significant impression this weapon made as one of the most successful means to capture cities. Nonetheless, the *Chronicle of the Morea* significantly makes the point that in spite of the impression they made and the psychological impact they must have had on the defenders, the trebuchets had no decisive influence on siege warfare in the Peloponnese. Instead, the surrender of cities is attributed to the destruction of their countryside, to agreements with the besiegers and to successful blockades that starved the besieged and compelled them to surrender. Similarly, in his description of the siege of Madytos by the Catalan Grand Company, Ramon Muntaner reports that the Catalans used a trebuchet (*trabuc*). However, Muntaner indicates that the Catalan assault did not centre on this weapon. The siege of Madytos was another small-scale operation, like almost all of the fourteenth-century sieges described by our sources. According to Muntaner, the Catalan force was composed of eighty horsemen and 200 foot-soldiers who had to face 700 defenders. The Catalans carried out many night attacks using ten rope ladders to climb the walls. Eventually, as Muntaner comments, one afternoon the Catalans exploited the fact that the walls were not sufficiently manned due to the excessive summer heat. They managed to surprise

[110] See France, *Western Warfare*, 120.

[111] *Chronicle of the Morea*, vv. 1410–1417, 1480–1485, 2033–2048.

[112] *Chronicle of the Morea*, vv. 2920–2935; A. Ilieva, *Frankish Morea (1205–1262)* (Athens, 1991), 194–195.

the defenders. The Catalans sent 100 men to climb the walls and captured two towers, while another group of soldiers tried to break down the gate with axes.[113]

The impact of gunpowder on Byzantine warfare

While in the middle of the fourteenth century Byzantium was plunged into civil wars, Western Europe witnessed the earliest use of gunpowder in its artillery. By the end of the fourteenth century guns were able to breach fortifications and in the early fifteenth century the adoption of smaller cannons and improvements in their accuracy increased their destructive capacity and prompted significant changes in the construction of fortifications.[114] Knowledge of the new technology spread from Western Europe to the Balkans and affected the Adriatic as early as the mid-fourteenth century.[115] It has been suggested that the earliest possible reference to the use of gunpowder by the Byzantines involves John VII's siege of the fortress of the Golden Gate in Constantinople, where his grandfather John V had taken refuge in 1390, following the coup by the former. This assumption relies on translations of the account of the Russian traveler Ignatius of Smolensk. While Kolias is cautious of an early twentieth-century German translation of Ignatius, in his analysis of the original Russian text, Bartusis doubts whether this siege involved the use of firearms.[116] Consequently, it is almost certain that by the early fifteenth century the Byzantines had not started to use gunpowder.

Moreover, the limited available information suggests that during the late fourteenth and early fifteenth centuries no major changes occurred in the siege warfare in Byzantium and in the wider world of

---

[113] Muntaner, II, 461–462.

[114] K. De Vries, *Medieval Military Technology* (New York, 1992), 144–148; idem, "The Impact of Gunpowder Weaponry on Siege Warfare in the Hundred Years War," in I.A. Corfis - M. Wolff (eds.), *The Medieval City under Siege* (Woodbridge, 1995), 227.

[115] D. Petrović, "Firearms in the Balkans on the Eve and After the Ottoman Conquests of the Fourteenth and Fifteenth centuries," in V.J. Parry - M.E. Yapp (eds.), *War, Technology and Society in the Middle East* (London, 1975), 170–173; G. Ágoston, *Guns for the Sultan. Military Power and the Weapons Industry in the Ottoman Empire* (Cambridge, 2005), 17.

[116] G. Majeska, *Russian Travelers to Constantinople in the Fourteenth and Fifteenth Centuries* (Washington, 1984), 102–103; Kolias, "Ανταρσία," 49; Bartusis, *The Late Byzantine Army*, 335–336.

Frankish Greece and the south Balkans. The *Chronicle of the Toccos* shows that the dominant form of warfare at this time was short sieges which could last as little as one or two days. They involved small armies which lacked the resources to train in the use of gunpowder artillery. The destruction of the economic basis of the enemy through raids and the seizing of forts through trickery (*klepsia*) are pointed by the anonymous chronicler as the main means of capturing fortified places. For instance, after destroying the countryside around Vonitsa the troops of the Spatas tried to capture it by using ladders at night. Similarly, the forts of Dragamesto, Aetos and Varnakas were captured by trickery when the soldiers of Carlo Tocco carried out a night attack using ladders.[117] In 1417/1418, John VIII, after conducting large-scale raids, captured fortresses which were in the hands of the prince of Achaia.[118] Describing the preparations of the Byzantine army in the Peloponnese to attack Clarenza in 1422 the *Chronicle of the Toccos* states that the Byzantines were prepared to come to Clarenza to blockade it, to destroy the forts around it, to disrupt its grain supplies and devastate the entire area.[119] Moreover, the same chronicle does not omit to mention the small size of the forces involved in these operations and relates that occasionally fortifications were captured by small groups of selected and experienced soldiers and because they were defended by insufficient forces.[120]

It is worth noting that the *Chronicle of the Toccos* mentions the use of bombards (*loumbardes*). In the siege of the tower of Anatolikon in 1406 the army of the duke of Cephalonia used *loumbardes* and the anonymous author attributes to the captain Matteo Ladolfo from Naples the construction of an engine which was 'able to pierce through the tower.' Its shots smashed the walls and the tower was captured.[121] This was the only instance mentioned by the chronicle in which the *loumbarda* was used to breach fortifications.[122] The absence of an elaborate

[117] *Cronaca dei Tocco*, vv. 94–95, 282–296, 968–996.

[118] *Cronaca dei Tocco*, vv. 3508–3529.

[119] *Cronaca dei Tocco*, vv. 3868–3871. In 1421, Carlo Tocco bought Clarenza from the adventurer Franco Oliverio from Puglia who had seized it from the Latin principality of Achaia. See Zakythinos, *Despotate*, I, 184; Bon, *Morée*, 286–288.

[120] *Cronaca dei Tocco*, vv. 57–106, 282–296, 775–795, 2510–2522.

[121] *Cronaca dei Tocco*, vv. 391–395.

[122] According to the same source *loumbardes* were part of the defense system of the fortress of Riniasa (Thomokastro) and were used as means of communication during military operations: *Cronaca dei Tocco*, vv. 2777, 2817, 3774–3775.

description prevents us from reaching safe conclusions regarding the form and destructive effect of the *loumbardes*. Nonetheless, it is evident from the account of the *chronicle of the Toccos* that their impact on military operations was rather limited. Moreover, the anonymous author makes a single reference to a trebuchet (*tetraboutseto*) used in 1403 by the troops of the Albanian chieftain Muriki Spata to capture the fort of Katoche in Epiros.[123]

The descriptions of sieges which are found in the chronicle of the Toccos are not much different from those provided by Sphrantzes. Sphrantzes' account of the siege of Patras by the Byzantines in 1429 makes no reference to siege engines and gunpowder. Instead, the Byzantines blockaded the city and destroyed the fields around it, while the defenders relied on the accuracy of their crossbowmen. Eventually, Patras surrendered to the Byzantines because its food supplies were exhausted.[124] Moreover, in his description of the siege and capture of Thasos by Manuel II in 1414, the author of the satirical work known as *Mazaris' journey to Hades* relates that the siege lasted three months and that Manuel II used stone-throwers (*petrovola mechanemata*). He makes no reference to gunpowder.[125]

Furthermore, until the end of the fourteenth century the capture of Byzantine cities and fortresses by the Ottomans in Europe was usually the outcome either of trickery or of agreements that guaranteed the lives of the inhabitants, saved the town from plunder and the besiegers from the expense of a costly siege operation. According to Chalkokondyles, in the 1360s the Ottomans captured Adrianople by surprising its inhabitants. The Ottomans realised that a young man was moving in and out of the city through an opening in the walls and they followed him.[126] According to Aşık Paşazade, following the capture of Pythion in ca. 1358, the Ottoman march lord Hacı İlbey seized Didymoteichon in ca. 1360 after ambushing its governor.[127] In

[123] *Cronaca dei Tocco*, vv. 391–395, 2497, 2777–2778, 3774–3775.

[124] Sphrantzes, 38–44, 58–60; *Kleinchroniken*, I, 236.

[125] *Mazaris*, 81, 83–85.

[126] Chalkokondyles, I, 28–29. For the Ottoman sources and the debate over the date of the Ottoman conquest of Adrianople see I. Beldiceanu-Steinherr, "La conquête d' Adrianople par les Turcs: La penetration Turque en Thrace et la valeur des chroniques Ottomans," *TM* 1 (1965) 439–461; H. İnalcık, "The Conquest of Edirne (1361)," *Archivum Ottomanicum* 3 (1971), 185–210; E. Zachariadou, "The conquest of Adrianople by the Turks," *Studi Veneziani* 12 (1970) 211–217.

[127] See Lowry, *Ottoman Balkans*, 20–21.

1387, Thessalonica surrendered to the Ottomans because of the activities on the part of the population which could not afford the cost of its defence.[128] Moreover, in the siege of Constantinople, which lasted from 1396 until 1402, the Ottoman army did not use gunpowder. Doukas relates that Bayezid I did not even use any siege engines and instead simply blockaded Constantinople to deprive it of its food supply although he is contradicted on this point by the Castilian ambassador, Ruy González de Clavijo, who was much closer to these events and by an anonymous contemporary source. Nonetheless, it is clear from Ottoman sources that even if Bayezid did have conventional siege engines, he did not deploy gunpowder weapons.[129] In 1392, the Genoese of Galata acquired more than sixty bombards and it is likely that some of them were used against the Ottomans. However, they were ineffective.[130] Therefore, blockade and attrition remained the staples of siege warfare throughout the period under discussion. While it is certain that the use of artillery, siege towers and the mining of the walls could allow storming forces to exploit the breaching of the walls and speed up the surrender of the besieged, the rather limited resources of the attackers made tight blockading preferable to assaults.

More convincing evidence regarding the use of gunpowder by the Byzantines and the Ottomans is provided by the sources of the Ottoman siege of Constantinople in 1422. Despite listing the many siege engines used by the Ottomans, who were led by Murad II, Kananos and Chalkokondyles remark that the Ottomans based their attack on cannons. However, their shots in the section of the wall between the Golden Gate and the Gate of Romanos did not breach the Byzantine walls.[131] Kananos' statement that the Ottomans built barricades in order to defend the city against the bows and crossbows of the Byzantines and the 'stones of the bombards' (*voumbardes*) leads to the conclusion that the Byzantines were using firearms.[132] The siege of 1422

[128] Dennis, *The Reign of Manuel II*, 77.

[129] Doukas, 79; Ruy González de Clavijo, *Embassy to Tamerlane*. tr. Guy Le Strange (London, 1928), 89; K.Heywood, "Notes on the Production of Fifteenth-Century Ottoman Cannon," 3. *Proceedings of the International Symposium on Islam and Science*, (Islamabad, 1981) reprinted in K. Heywood, *Writing Ottoman History* (Aldershot, 2002); P. Gautier, (ed.), "Un récit inédit sur la siege de Constantinople par les Turcs (1394–1402)," *REB* 23 (1965), 106. For the impact of the siege on the population of Constantinople see Necipoğlu, *Between the Ottomans and the Latins*, 149–183.

[130] See Bartusis, *The Late Byzantine Army*, 336.

[131] Chalkokondyles, II, 10; Kananos, 7.

[132] Kananos, 6; Bartusis, *The Late Byzantine Army*, 338.

is considered the earliest reference to Ottoman cannons. Nonetheless, it has been argued that there were Ottoman cannoneers during the reign of Mehmed I (1413–1421).[133] In another instance indicating the use of gunpowder by the Byzantines, discussing a clash between the Byzantines and the Genoese of Galata over trade in Pera in 1434, Chalkokondyles comments that both sides used cannons (*televola*).[134]

In the 1440s the tactical significance of artillery increased as a result of the wars that the Ottomans fought against the Hungarians and it is likely that Murad II established a separate artillery corps as part of the salaried standing army of the Ottomans.[135] The Ottomans used cannons in their attack against the wall of Hexamilion in 1446, which was led by Murad II. Due to the enormous size of these weapons, the Ottomans transported the raw materials with them and cast the cannons *in situ*.[136] Nonetheless, Chalkokondyles indicates that, although the cannons prevented the defenders from staying very long in their posts, they did not breach the walls. Instead, the Ottoman victory is attributed to the successful use of ladders and to the inexperience of the defenders, who fled once they saw the enemy climbing the wall.[137] That the Ottoman cannons in the 1440s were not particularly effective is indicated by the siege of Belgrade when the use of gunpowder artillery failed to force the city to submit, despite the fact that its fortifications were designed for cold steel warfare.[138]

The role of the Ottoman firearms and cannons in the fall of Constantinople in 1453 is well-known and it would be superfluous to discuss it here.[139] It suffices to say that the siege of Constantinople shows that the Ottomans made far better use of gunpowder artillery than the Byzantines, who seem to have lacked comprehensive knowledge of its deployment. This lack of understanding may have

---

[133] Petrović, "Firearms in the Balkans,"190; Ágoston, *Guns for the Sultan*, 20.

[134] Chalkokondyles, II, 61–62. For this conflict see Necipoğlu, *Between the Ottomans and the Latins*, 190–191.

[135] Ágoston, *Guns for the Sultan*, 17, 28.

[136] Heywood, "Ottoman Cannon," 5; Ágoston, *Guns for the Sultan*, 94.

[137] Chalkokondyles, II, 114–116.

[138] M. Popović, "The Fortress of Belgrade," in S. Ćurčić – E. Hadjitryphonos (eds.), *Secular Medieval Architecture in the Balkans, 1300–1500, and its Preservation* (Thessalonica, 1997), 128–131; idem, *Beogradska tvrdjava* [*The Fortress of Belgrade*] (Belgrade, 2006) 2nd edition with a summary in English, 326.

[139] See Bartusis, *The Late Byzantine Army*, 338; De Vries, "Gunpowder Weapons at the siege of Constantinople in 1453," in L. Yaakov (ed.), *War, Armies and Soldiery in the Eastern Mediterranean* (Leiden, 1986), 343–362.

caused friction among the leaders of the besieged. If we believe the testimony of Leonard of Chios, Loukas Notaras refused the request of Giovanni Giustiniani (the head of the Genoese reinforcements sent to Constantinople) to use the cannons that belonged to the city in the section that he commanded.[140] An instance that indicates insufficient technical knowledge on the part of the Byzantines, as well as the panic of those under siege, is the information provided by Chalkokondyles that when the largest cannon used by the Byzantines exploded, the gunner was accused of having received bribes from the enemy.[141]

A very important question is whether the repairs to the fortifications of Constantinople in the fifteenth century corresponded to the technological changes in the art of war. In their study of the walls of Constantinople, Foss and Winfield conclude that the Palaiologoi constructed higher and stronger towers which would protect the city against artillery and trebuchets. They also suggest that circular loopholes, appropriate for the use of guns, were built in several towers and date from the Palaiologan period.[142] Moreover, among the Italian defenders of Constantinople there were hand gunners and Doukas relates that the Byzantines used small canons. The cannon balls were as small as walnuts. However, such was their velocity that with one shot it was possible to kill two or three soldiers.[143] It could be argued that the testimony of Doukas is not reliable because he was not an eyewitness and his account of the Byzantine guns is almost identical to his description of the canons used by the defenders of Belgrade in 1440 during Murad II's unsuccessful siege.[144]

Nevertheless, the construction of high towers is not the ideal way to defend against gunpowder weapons. Furthermore, Leonard of Chios and Chalkokondyles relate that the fortifications of Constantinople were unable to support cannons. According to Leonard of Chios, in the siege of 1453 the largest cannon of the Byzantines did not fire a single shot, out of fear that it would damage the walls, while Chalkokondyles informs us that the shots of the Byzantine artillery shook the walls of

[140] Pertusi, *La Caduta*, 153; Jones, *The Siege*, 29.

[141] Chalkokondyles, II, 154.

[142] Foss – Winfield, *Byzantine Fortifications*, 52. For a survey of the walls of Constantinople see N. Asutay – Effenberg, *Die Landmauer von Konstantinopel-Istanbul. Historisch-Topographische und baugeschichtliche Untersuchungen* (Berlin, 2007), passim.

[143] Doukas, 331.

[144] Doukas, 263–265; Bartusis, *The Late Byzantine Army*, 340.

Constantinople causing more damage to them than to the enemy.[145] These testimonies indicate that the Byzantines failed to assess the impact their cannons would make on the walls, before they used them in a war situation. The Byzantines understood the importance of the new military technology and supplied themselves with guns and cannons. However, in this early stage they seem to have been unaware of the full extent of the changes needed to the walls of the city and were unable to exploit the full potential of gunpowder weaponry. It needed time to fully assess and implement the necessary changes in military architecture in order to improve the effectiveness of the new weapons. Moreover, on the eve of the fall of Constantinople the Byzantines were unable to find and too poor to hire specialists who could bring to Byzantium new ideas about fortification architecture.[146]

However, the system commonly used by the later Byzantines in building fortifications did not differ from that used in late antiquity and even fortresses, such as Pythion, the design of which indicates the use of highly-skilled workers and an awareness of the development of military architecture in thirteenth and early-fourteenth century Western Europe, were built with cold steel warfare in mind.[147] It is certain that the lack of resources explains the inability of the Byzantines to follow the technological developments that occurred in Western Europe, where the late fourteenth century witnessed the introduction of gunports.[148] Moreover, the limited use of gunpowder technology by the Byzantines and their failure to adapt their fortifications to the needs of firearms warfare should be examined in the context of the developments in siege warfare in the Balkans. While cannons were used on the Adriatic coast as early as the middle of the fourteenth

---

[145] Pertusi, *La Caduta*, 132; Jones, *The Siege*, 16; Chalkokondyles, II, 154.

[146] For the economic conditions in Constantinople in the siege of 1453 see Necipoğlu, *Between the Ottomans and the Latins*, 222–225; idem, "Social and Economic Conditions in Constantinople during Mehmed II's Siege," in T. Kiousopoulou (ed.), *Η Άλωση της Κωνσταντινούπολης και η μετάβαση από τους μεσαιωνικούς στους νεώτερους χρόνους* [The capture of Constantinople and the Transition from the Medieval to the Modern Times] (Herakleion, 2005), 75–86.

[147] Ćurčić, *Architecture in the Balkans*, 512–518. It has been argued that the inherent conservatism of Byzantine architecture was the result of the fact that architecture in the later periods was based on practice rather than on theory: R. Ousterhout, *Master Builders of Byzantium* (Philadelphia, 2008), 44.

[148] De Vries, "The Impact of Gunpowder," 230, 246; idem "Gunpowder Weaponry," 233–236.

century, the use of gunpowder weapons went on to influence the rest of the Balkans in the course of the fifteenth century.

Moreover, the ability to produce cannons did not coincide with the remodeling of fortifications in accordance with the demands of gunpowder warfare. For instance, while Dubrovnik a few decades after 1350 became the centre of production and exporter of gunpowder weapons, the modification of its walls and adaptation to the impact of firearms was a gradual process and it was not until 1461 that a Florentine architect, Michelozzo di Bartolomeo, who had been invited and employed by the city of Dubrovnik, introduced new ideas in fortification architecture. The increasing menace of Ottoman invasion led to the decision not to hire a local master, but to employ an eminent Italian architect-engineer.[149] The fortifications of Smederevo, which were built in 1430 by the Serbian prince George Branković (1427–1456) under the supervision of his Byzantine brother-in-law, George Palaiologos Kantakouzenos, followed to a large extent Byzantine principles of cold steel warfare.[150] It was after the 1450s that Smederevo's fortifications were upgraded to take account of the new technological developments.[151] Similarly, the fortifications constructed by the Serbian prince Stefan Lazarević (1402–1427) in Belgrade were based on older developments of Serbian military architecture and it was under Hungarian rule in the 1440s that new fortifications, including lower round towers with thicker walls and emplacements for cannons, were built and were adapted to the needs of firearm warfare. Nonetheless, the Ottoman siege of the city in 1456, although a failure, revealed the imperfect techniques in artillery warfare on the part of the defenders and called for further modifications.[152] That the fortifications of Dubrovnik were adapted to the needs of firearms warfare over a century after the city started producing cannons, that the fortifications of Belgrade needed further remodeling after the siege of 1456 and that in 1453 the defenders of Constantinople did not know that the walls

[149] H. McNeal-Caplow, "Michelozzo at Ragusa: New Documents and Revaluations," *Journal of the Society of Architectural Historians* 31/2 (1972), 110–111; Petrović, "Firearms," 173–182; Ćurčić, "Architecture in the Age of Insecurity," 30–31.

[150] George Palaiologos Kantakouzenos was a grandson of John VI Kantakouzenos' son, Mathew. See Nicol, *The Family of Kantakouzenos*, 176–179.

[151] Ćurčić, "Architecture in the Age of Insecurity," 33–34; N. Jocović – J. Nešković, "Fortifications of Smederevo," in S. Ćurčić – E. Hadjitryphonos (eds.), *Secular Medieval Architecture in the Balkans, 1300–1500, and its Preservation* (Thessalonica, 1997), 133.

[152] Popović, "The Fortress of Belgrade," 129; idem, *Beogradska tvrdjana*, 325.

of their city were unsuitable for the placement of cannons, show that in the middle of the fifteenth century the armies of the small polities in the Balkans had not developed any mastery in using gunpowder weapons. They were still experimenting and learning the use of new technology. Therefore, they were unable to fully exploit its potential. It should be added that the Ottomans, who in the fifteenth century were the strongest power in the region and were able to acquire knowledgeable engineers from abroad and highly skilled builders, made limited use of gunpowder and even they, as has been noted, did not develop a mastery of military architecture. This reluctance to make extensive use of gunpowder artillery must have followed a decision of the military leadership that cannons did not serve the military objectives of the Ottoman state.[153]

Consequently, as was the case in the other Balkan states, in Byzantium, the decision to adopt or reject the use of gunpowder weapons depended on the priorities of the throne and was influenced by the economic capabilities of the late Byzantine state, the mentality of the ruling elite and its military objectives. The effective use of cannon required the storage and preservation of adequate quantities of gunpowder and the employment or training of experts in pyrotechnical matters and metallurgy. The procurement of raw materials for the large-scale construction of gunpowder weapons and the creation of specialised troops trained to use the new weapons would have been an impossible task for the fifteenth-century Byzantine emperors and semi-independent despots, given the impoverishment of the Byzantine state and the shortage of manpower. Furthermore, it is safe to infer that the social background and military tradition of the Byzantine military elite favoured a cautious approach towards all new weapons. Byzantine generals assumed military command because they conformed to social criteria and training remained largely a personal issue. The Byzantine aristocratic elite expressed its class exclusiveness as heavy cavalrymen and its members were trained in a different kind of war. It is worth noting that initially gunpowder weapons were received equally suspiciously by Western European knights who were accustomed to a different kind of fighting and saw their dominant role in warfare being threatened by the introduction of cannons. In a similar manner, before the tactical significance of gunpowder weapons

[153] Ćurčić, "Architecture in the Age of Insecurity," 36; Ágoston, *Guns for the Sultan*, 17–21.

became clear, the Ottoman *sipahis* saw the use of firearms as unworthy of their status.[154]

From the purely tactical point of view, the transport of cannons could seriously slow down the rate of march of campaigning armies and the small political entities of the period could not afford to support the logistical needs of long campaigns. Warfare in the south Balkans and Frankish Greece of the fifteenth century was characterised by short and swift campaigns conducted by small armies and often success relied on the element of surprise. Similarly, most of the sieges of the period were short operations concentrated on the destruction of the enemy's hinterland and on trickery. Therefore, heavy siege equipment could significantly delay, as well as advertise, the advance of campaigning forces. In addition, as was the case with every medieval commanders and engineers, the Byzantine military leaders could not always know in advance the kind and number of siege engines they might be able to use in a siege. This required an accurate knowledge of the enemy fortifications and the topography of the area, which dictated the choice of devices to employ.[155] Not all Byzantine commanders possessed such knowledge and therefore improvisation played a key role in the conduct of many siege operations. In 1323, the leaders of the Byzantine army that besieged Phillippoupolis had not planned in advance to construct a siege tower.

In his proposals for the reorganisation of the defence of the Peloponnese in 1444/1446, Cardinal Bessarion emphasised the value of developing professional military forces. After pointing out the need to maintain a well-trained and disciplined army, he exhorts the Byzantine leaders to send young men to Italy to learn sciences. They should receive training in engineering, mining and metallurgy and be taught the construction of arms and shipbuilding, which were essential for the development of a professional and effective army.[156] While the Ottomans under Mehmed II were able to recruit Western European specialists who enabled them to build fortresses using advanced military theories of fortifications, such as Kilid- ül- Bahir in Gallipoli (Gelibolu), it is obvious that Bessarion's proposals never materialised.[157]

---

[154] De Vries, *Military Technology*, 159; Ágoston, *Guns for the Sultan*, 57.

[155] See Rogers, *Latin Siege*, 9.

[156] *PP*, IV, 42–43.

[157] B. Özgüven, "Fortress of Kilid-ül-Bahir," in S. Ćurčić – E. Hadjitryphonos (eds.), *Secular Medieval Architecture in the Balkans, 1300–1500, and its Preservation* (Thessalonica, 1997), 139–143; Ćurčić, *Architecture in the Balkans*, 568–570.

CHAPTER SEVEN

TACTICS

Late Byzantine cavalry and field tactics against Western Europeans

The late Byzantine cavalry had to respond successfully to its two main opponents, the Western Europeans and the Turks, who used different tactics and weapons. From a very early time Byzantine generals had realised the benefit of understanding the setbacks of the enemy tactics and of adjusting the deployment of their troops accordingly.[1] The sixth century *Strategikon* of Maurice discusses in detail how other people fight, and by what methods they could be dealt effectively by the Byzantines. Leo's *Taktika* was to a great extent 'copied' from the *Strategikon*, but the copied sections often reveal editorial 'updating.'[2] Both treatises state that the Franks are rash and daring in fighting but they easily lose their discipline. They suggest that the Byzantine generals should avoid pitched battles against them. Instead, they should deploy stratagems and ambushes, to try to prolong their encounter, exhausting them. Furthermore, they comment that it is safe to fight against them in difficult-to-cross areas and in woodland.[3]

Although narrative histories seldom provide detailed tactical information, late Byzantine historians express similar views. However, they were influenced less by middle Byzantine military treatises, and more by contemporary realities. Between the eleventh and fifteenth centuries, the heavy armoured cavalry played a decisive role in Western European warfare. Western European tactics relied on mass and close order and their aim was to bring the confrontation to close quarters where they could take advantage of their heavy equipment and armour. The main tactics they followed were charges and perhaps feigned retreats. To be successful these tactics had to be well timed and delivered in a disciplined, close-order fashion, backed up by infantry.[4]

---

[1] McGeer, *Sowing the Dragon's Teeth*, 225.

[2] Haldon, *Warfare State and Society*, 315.

[3] *Strategikon*, XI. 1–XI. 3; *Taktika*, XVIII. 88–92.

[4] See Verbruggen, *The Art of Warfare*, 103–104; France (*Western Warfare*, 159) states that, although by the thirteenth century the mass cavalry charge became decisive, modern writers are obsessed with the shock role of the knight and that the theory

Moreover, as heavier armour developed, new breeds of horses of greater size and weight were developed, in response to the demands of shock combat and the burden of armour.[5] The late Byzantines were aware of the above-mentioned developments and point out the heavy armour used by the Western European armies. Describing the battle of Adrianople (1205) where the Latins of Constantinople were defeated by the Bulgarians and the Cumans, Akropolites states that as it was the custom for the Latins to ride towering horses and to wear full body armour, they moved with difficulty against their enemies.[6] In his account of the battle of Pelagonia he remarks, that Manfred, the Hohenstaufen king of Sicily, sent to Michael of Epiros 400 knights whom he describes similarly, emphasising their full body armour and the large size of their horses.[7] Gregoras relates that the army which was sent in April 1281 to relieve the besieged Bellagrada in Albania had orders to avoid engaging in pitched battles against the Angevins, since they had very good weapons and were heavily armoured.[8] Pachymeres also makes a reference to the large size and heavy armour of horses of the forces of the general of Charles I of Anjou in Epiros and Albania, Hugue Rousseau of Sully.[9]

The Byzantine heavy cavalry was less armoured than the Western European, and there are indications that the Byzantine horses were smaller than Western European ones.[10] The Byzantine generals realised that the best way to deal with the Frankish heavy cavalry was the deployment of light cavalry and infantry troops, and the avoidance of pitched battles. The sources show that the Byzantine military leaders deployed stratagems trying to take advantage of the unfamiliarity of the Frankish heavy cavalry with the terrain, bearing in mind that the Balkan and Asia Minor landscape was not favourable for the

of shock tactics grossly overstates the value of the horseman as a 'weapons system'. Keen (*Chivalry*, 23) states that since the eleventh century the charge of heavy cavalry men holding their lances in the couched position could decide the outcome of the battle.

[5] A. Hyland, *The Medieval Warhorse from Byzantium to the Crusades* (London, 1994), 9.

[6] Akropolites, I, 22.

[7] Akropolites, I, 168; Pachymeres (I, 117) states that Manfred sent 3000 (a grossly exaggerated figure) German knights (*kavallarioi*). Sanudo, (115) gives a figure identical to Akropolites'.

[8] Gregoras, I, 55–56, 147.

[9] Pachymeres, II, 647.

[10] Theodore Palaiologos, *Enseignements*, 58.

deployment of heavy cavalry tactics. They were also aware of the fact that the heavily armoured cavalry charges, if defended properly, were exhausting and the riders grew weary. The aim of the Byzantines was to break the discipline of the Latins by cutting off their supplies and breaking their co-ordination.

The elite Byzantine soldiers, the *pronoia* holders, or at least the wealthier of them, as well as the members of the aristocracy and probably many Latin mercenaries were heavy cavalry soldiers. Akropolites, blames Manuel Laskaris and Constantine Margarites for imprudently attacking a Cuman raiding force close to Didymoteichon (1256), saying that the Byzantines were accustomed to being fully armoured (*cataphracts*) with heavy weapons.[11] This means that the soldiers of Didymoteichon were heavily armed. In 1255, Theodore II in his campaign against the Bulgarians in Serres was leading a force of heavy cavalry.[12] In the battle of Pelagonia (1259) the Byzantine army included heavy cavalry troops.[13] It should be noted, however, that in none of the above-mentioned examples were heavy cavalry soldiers an effective force. In the first instance, they were misused. In Didymoteichon they fell into the trap of the feigned retreats of the mobile Cuman cavalry, which exhausted them. In Serres they were not used, because the geographical setting of the battle did not favour the deployment of cavalry, and in the battle of Pelagonia John Palaiologos did not deploy them for tactical reasons. It had been decided that the Turkish and Cuman light cavalry archers, as well as Byzantine foot archers were more suitable to deal with the Latin heavy cavalry.

Furthermore, the Byzantine troops involved in the warfare against the Latins in the Peloponnese were predominantly light cavalry. In 1262, Constantine Palaiologos had under his command troops from Asia Minor, and Turkish mercenaries. Pachymeres states that this campaigning force did not include 'Italians,' because it was not proper to have 'Italians' fight against 'Italians.'[14] It is tempting to conclude that Pachymeres implies that the Byzantines avoided using mercenary soldiers of similar cultural background with the enemy. However, this was not the only thing Pachymeres intended by this statement. Late Byzantine historians, and particularly Pachymeres, referred to

[11] Akropolites, I, 126.
[12] Akropolites, I, 147.
[13] Akropolites, I, 168.
[14] Pachymeres, I, 273.

all the westerners as 'Italians' and 'Latins.' For instance, they call the Catalans 'Italians' and the Germans 'Latins.'[15] The Western mercenaries employed by Byzantium were not exclusively Italians, while most of the Latin knights in the Peloponnese were predominately 'French.' Beyond that, it is premature to talk about national conscience of mercenaries in the late Byzantine period, and it is highly unlikely that mercenary soldiers in the period under discussion would have had strong national convictions. Rather, Pachymeres implies that the Byzantine military command thought that tactically it was not useful to deploy heavy cavalry troops against the Frankish cavalry of the Peloponnese, which would have been superior to the heavy cavalry employed by the Byzantines. Instead, the Byzantines would use light cavalry and guerrilla warfare tactics as they had done in Pelagonia. Similarly, it seems that the Byzantine heavy cavalry did not play any significant role in the Byzantine operations against the Latins in Epiros in the 1280s.

The tactical aim of the late Byzantines was to force the Latins to lose their discipline and break their ranks. Gregoras states that,

> When the Italian nation fights in an orderly way it is like a fortified and impregnable wall. However, if they break their ranks slightly, there is nothing to prevent their enemies from capturing them as prisoners.[16]

The most effective way to make the Western European heavy cavalry troops lose their discipline was to follow guerrilla warfare tactics. Byzantine commanders realised that the best means to achieve their tactical aims against the Latins was the use of Turkish and Cuman cavalry archers, whose tactics relied on speed, mobility, surprise and the use of archery to avoid pitched battles. They rest on ambush and feigned retreats aiming at the disruption of the enemy's order and at the exhaustion of the heavily armoured Latin cavalry. Warriors like the Cumans received training in riding horses and in archery from a young age and that their logistical demands were less than the needs of the Western cavalrymen.[17]

---

[15] For the use of these terms in late Byzantine sources see P. Karlin – Hayter, "Notes sur le ΛΑΤΙΝΙΚΟΝ dans l' armée et les historiens de Nicée," *BF* 4 (1972), 142–150; A. Laiou, "Italy and the Italians in the Political Geography of the Byzantines (14th Century)," *DOP* 49 (1995), 74–76.

[16] Gregoras, I, 147.

[17] D. Sinor, "The Inner Asian Warriors," *Journal of the American Oriental Society*, 101 (1981), 133–144; N. Di Cosmo, "Inner Asian Ways of Warfare in Historical Perspective," in idem (ed.), *Warfare in Inner Asian History* (Leiden, 2002), 3; Hyland,

A number of battles fought in the period under discussion confirm the above-mentioned statements that the Byzantines employed light cavalrymen and guerrilla battlefield tactics against their Latin opponents. The first major engagement illustrating the effectiveness of light cavalry against the heavily armoured Frankish cavalry which, however, did not involve Byzantine troops, was the well-documented battle of Adrianople where in 1205 the Bulgarians and the Cumans defeated the Latins of Constantinople. According to Gregoras, at the initial stage of the battle the Bulgarians and Cumans feigned retreat. After a long pursuit by the Latins, the Cumans returned unexpectedly and by launching their arrows and javelins, defeated the Latins, who were exhausted by the long chase.[18] Very similar is the account of the *Chronicle of the Morea*, which attributes the Latin defeat to the fact that the Franks were not used to this form of warfare, and were provoked into pursuing the enemy, who managed by their missiles to kill the horses of the Franks.[19] This can be seen as an indication that Cuman arrows could not penetrate the armour of European knights, but they could kill their mounts. William of Villehardouin, who as an eyewitness must have been the most reliable source on this battle, states that the battle lasted two days, and although the Latins held a war council at night and decided that it was unwise to pursue for long the light enemy forces, the very next day they repeated the same mistake. Choniates adds that the Cumans and Bulgarians took advantage of the unfamiliarity of the Latins with the area and ambushed them.[20]

The Byzantines adopted similar battlefield tactics. In 1211, the Nicaean army under Theodore I encountered the army of the Latin empire of Constantinople. According to the letter of Henry of Flanders which was addressed to the pope Innocent III, Theodore I ordered his troops on a hill and his aim was to draw the Latins over to rough ground where a series of ambuscades had been prepared. However,

---

The Medieval Warhorse, 126; Jackson, *Mongols and the West*, 32; Robert de Clari, (64) makes a very interesting description of the battlefield tactics of the Cumans, which also illustrates the perceptions the Latin knights had of them. Clari depicts the Cumans as savages, who live in tents and do not cultivate the land. Each Cuman has at least ten or twelve well-trained horses. The cavalrymen have their horses carrying their fodder when on the march and therefore, they feed while following their master. The cavalry does not stop at night; the Cumans ride so hard that in one day and one night they cover six or seven or eight days' journey.

[18] Gregoras, I, 15.

[19] *Chronicle of the Morea*, vv. 1114–1161.

[20] Villehardouin, 169–172; Choniates, *Historia*, 616.

the Latins discovered the Byzantine plan.[21] The Byzantines were more successful in the battles of Pelagonia (1259) and in Bellagrada (1281). The Byzantine victory in Pelagonia was the result of the mobility of the Turkish and Cuman cavalry archers and their ability to hit the enemy from a distance. They were able to ambush the enemy cavalry troops, repeatedly preventing them from reaching water resources for their horses. They also conducted many raids on the enemy baggage train.[22] According to the *Chronicle of the Morea*, John Palaiologos would never have won the battle of Pelagonia without the aid of archery.[23] The Byzantine troops sent by Michael VIII to relieve Bellagrada from the siege of the forces of Charles of Anjou were ordered not to provoke the Angevin forces to field battle. The siege was lifted when the enemy commander, Hugue Rousseau of Sully was ambushed and captured by Turkish mercenaries.[24] Moreover, most probably the Byzantine armies involved in the operations in the 1280s early 1290s against the Angevin possessions in Epiros and Albania included substantial numbers of Turkish mercenaries. However, very little is known about the nature of these operations.

As far as the warfare between the Byzantines and the Franks in the Peloponnese is concerned, the *Chronicle of the Morea*, despite its inaccuracies and prejudices, reflects vividly the differences in battlefield tactics and attitudes towards warfare adopted by the Byzantines and their enemies in the Peloponnese. As has been discussed in a different context, this source is negative towards the Byzantines and insists that they avoided fighting the Franks in the open field stating that the Byzantines 'would not fight the Franks with lances but with bows,' to demonstrate that the Byzantine soldiers were cowards in contrast to the brave Frankish warriors, who sought to fight in close quarters. However, such views about bravery on the battlefield have little to do with the realities of warfare. As the author of the *Chronicle* admits, 'stratagems and cunning tactics win over bravery.'[25] He also relates that the aim of the cavalry archers was to shoot and kill the horses of the knights and not the knights themselves and attributes the

---

[21] Prinzing, "Brief," 414–416.

[22] Akropolites, I, 168–169; Pachymeres, I, 121. Although, his version differs, he points to the contribution of the mobile Turkish and Cuman cavalry.

[23] *Chronicle of the Morea*, vv. 4920–4921.

[24] Pachymeres, II, 645; Gregoras, I, 147; Sanudo, (145) states that they were Turks.

[25] *Chronicle of the Morea*, vv. 4907, 4964.

Byzantine defeat in Prinitza in 1262 not to the effectiveness of the tactics of the Latins, but to the fact that the Byzantines underestimated the small number of Frankish cavalry and misused the Turkish cavalry archers.[26] The *Chronicle of the Morea* indicates that in the Peloponnese the Byzantines conducted raids the aim of which was to damage the morale and the resources of the Latins; they avoided pitched battles. The Franks sought pitched battles. However, they were extremely cautious when they had to deal with effective light cavalry and archery.[27] For instance, in 1270 Michael VIII sent an army in Monemvasia which was composed of Cumans, Turks and Byzantines. The Latins under William II decided not to confront the Byzantines. Instead, they reinforced the defences of their fortifications. In 1272, Charles I of Anjou sent William II of Villehardouin reinforcements to fight against the Byzantines. In a war council the leaders of the Latins decided to encounter the Byzantines in a pitched battle. Wishing to entice the Byzantine army to fight, the Latins plundered Byzantine possessions in the Peloponnese. However, the Byzantines avoided battle. Instead, they laid ambushes in inaccessible areas. Realising that the Byzantines had no intention to fight and that it was too risky to march through the woodland and hills, the Latins retreated. They did not wish to become easy targets for the Turkish and Cuman archers.[28]

Battlefield Tactics against the Turks

The study of Byzantine-Turkish warfare during the period under discussion is rather interesting since it was the Turkish military strength which brought about the end of the Byzantine empire. The most characteristic example reflecting the difficulties faced by the Byzantine army when confronted by the armies of the Turcoman principalities in this period is the battle of Pelekanos, in the gulf of Nikomedia.[29] This battle illustrates the problems for the Byzantines in confronting enemies who relied on light cavalry archers. It was fought in 1329 between the Byzantines, under the command of the emperor

[26] *Chronicle of the Morea*, vv. 4931–4944, 4967.

[27] *Chronicle of the Morea*, vv. 5084–5092.

[28] *Chronicle of the Morea*, vv. 6487–6720.

[29] On this battle see S. Kyriakidis, "Byzantine Responses to the Battlefield Tactics of the Turkoman Principalities: The Battle of Pelekanos," *BZ* 103 (2010), 83–97.

Andronikos III and the Ottomans, who were led by Orhan. According to a short chronicle, the Byzantine army departed to Mesothynia on 1 June 1329.[30] It arrived in Pelekanos, at the passage where the fortresses of Philokrene, Niketiatou, Dakivytza and Retzio were located.[31] Seeing that the Ottomans had already captured the surrounding hills and were aware of their moves, the Byzantines halted their march and decided to withdraw. Kantakouzenos relates that Orhan deployed his troops on high ground which was difficult to reach and adds that if the Ottomans had had no intention to fight and remained in their positions, the Byzantine army would have returned to Constantinople. He argues that it should not have been seen as an indication of cowardice that the Byzantines withdrew. Indeed, it was the reasonable thing to do, since it was impossible to fight the enemy.[32]

Eventually, the Byzantines ordered their line on flat terrain, which, according to Kantakouzenos, was suitable for heavy cavalry manoeuvres.[33] Orhan ordered his troops on a hill and placed part of his army in ambushes.[34] According to Kantakouzenos, Orhan started the battle by detaching 300 cavalry archers to entice the Byzantines into an undisciplined charge. Andronikos III sent out 300 cavalrymen to drive the Ottomans off. The emperor ordered them to advance cautiously and in close order against the Ottomans. They were also instructed not to pursue the enemy more than necessary because they would get exhausted and as a result they could break order and fall into ambushes.[35] Kantakouzenos relates that when they saw that their shots did not break the order of the Byzantines, the Ottomans retreated. He adds that the Ottomans carried out many similar attacks but all of them failed. However, the broken ground prevented Andronikos III from sending forward his whole army. It was impossible for the Byzantines to encircle the enemy.[36]

Kantakouzenos and Gregoras differ in the reaction they ascribe to the Ottoman tactics on the part of Andronikos III. Kantakouzenos implies that the Byzantine soldiers who were sent to force back the Ottoman cavalry archers owed their success to the emperor's orders.

[30] *Kleinchronken*, I, 78.
[31] For this fortresses see Foss, *Nikomedia*, 44–58.
[32] Kantakouzenos, I, 343.
[33] Kantakouzenos, I, 344.
[34] Kantakouzenos, I, 344–345.
[35] Kantakouzenos, I, 347.
[36] Kantakouzenos, I, 343–348.

Gregoras, in contrast, writes that it was a mistake to send troops against the Ottoman cavalry archers, whose feigned retreats aimed to draw the imperial forces over to rough ground where a series of ambuscades had been prepared. Gregoras also relates that Andronikos III saw this method of fighting by the Ottomans as evidence of cowardice and, encouraged by this belief, he sent Byzantine soldiers against them, contrary to the advice of the more experienced commanders around him. As a result, the Byzantines were constantly in a defensive position and exhausted themselves in the summer heat.[37] This account reflects not only Gregoras' negative view of Andronikos III, but also his different approach to the outcome of the battle. As the supreme commander of the Byzantine army and a close associate of the emperor, Kantakouzenos emphasises the success of the Byzantines in being able to force back the Ottoman cavalry archers without losing their order. As a critic of the generalship of Andronikos III, Gregoras sees the decision to send troops against the Ottoman provocation to attack as an illustration of Andronikos' misunderstanding of Ottoman battlefield tactics.

In the next stage of the battle, Orhan ordered a general assault against the Byzantines which was led by his brother Pazarlı. However, the Byzantines successfully defended themselves and carried out a counterattack, killing many enemies. Believing that it was too risky to advance any further, the Byzantines decided to withdraw and prepared to return to Constantinople, unless the Ottomans decided to fight on open ground.[38] Both Kantakouzenos and Gregoras point out that the lack of discipline and the false rumour that the emperor had been fatally wounded turned the Byzantine withdrawal into panicked flight. Eventually, the army was dispersed into the fortresses around Nikomedia and departed to Constantinople.[39]

It is interesting that in his account of the Byzantine withdrawal Kantakouzenos has included an incident which typifies the inability of the Byzantines to deal effectively with the Turkish cavalry archers. He relates that a certain Sevastopoulos, who was, 'a Bulgarian submitted to the emperor,' probably a local magnate or brigand, and had joined the campaign at the head of 300 poorly armed men, used poor

[37] Gregoras, I, 434.
[38] Kantakouzenos, I, 349–352; Gregoras, I, 435.
[39] Kantakouzenos, I, 362–363; Gregoras, I, 436.

quality horses and held the last position in the army, assembled his troops and some infantrymen from the rest of the army and hastily advanced against the Ottomans. Kantakouzenos comments that the troops of Sevastopoulos were defeated because they did exactly what he and Andronikos III had instructed the army not to do during the battle. As he writes, the Turks pretended that they were defeated and feigned retreat. Sevastopoulos' troops 'stupidly,' as Kantakouzenos calls it, pursued them. Then suddenly, the Ottomans counterattacked and put them to flight.[40]

The battle of Pelekanos as described by Gregoras and Kantakouzenos shows that the Byzantine military leadership possessed a good knowledge of the fighting methods of the Turks. Andronikos III's instructions to his soldiers, as described by Kantakouzenos, and Kantakouzenos' account of the engagement between the troops of Sevastopoulos and the Ottomans prove that the Byzantine generals were aware how much the Turkish armies relied on the mobility of light cavalry archers, on feigned retreats and ambushes, and on the avoidance of fighting at close quarters against heavily armoured cavalry. It is worth noting that Kantakouzenos' description of the fighting methods of the armies of the Turcoman principalities and his suggestions concerning the Byzantine responses are in accordance with the instructions provided by military treatises of earlier epochs, such as the *Strategikon* of Maurice, and the *Taktika*. According to the *Strategikon*, warriors such as the Turks and the Avars are well-trained as mounted archers. On campaign they are followed by numerous spare horses and they place a great many troops outside the line of the battle, in order to lay ambushes. Furthermore, the author of the *Strategikon* observes that, since the Avars and the Turks prefer to fight their enemies from a distance and to deploy ambushes and feigned retreats, Byzantine soldiers should never pursue them without taking precautions. He also relates that it is best to fight the Turks in the open field, using dense cavalry formations.[41] As has been shown above, this was the aim of the Byzantines at Pelekanos.

The author of the *Taktika* adopts the comments of the *Strategikon* and points out the reliance of the 'Saracens' on archery. He adds that the Byzantines should engage them at close quarters without delay.

[40] Kantakouzenos, I, 352–354.
[41] *Strategikon*, XI. 2.

By doing so, the Byzantine soldiers would avoid any damage from the enemy arrows and would take advantage of the enemy's lack of armour. The *Taktika* also comment that, because arrow shots damage the coherence and order of heavy cavalry, the Byzantines should engage mobile cavalry archers only on the plain.[42] Consequently, long before the fourteenth century the Byzantines were aware of the fighting methods of enemies who relied on the mobility of light cavalry archers and the examination of the battle of Pelekanos shows the continuity of the military practice of bringing the enemy to close action as swiftly as possible, if an army of mobile light cavalry archers was to be attacked.

Furthermore, the available source material suggests that, throughout the thirteenth and early fourteenth centuries, the armies of the Turcoman principalities in Asia Minor followed battlefield tactics which were similar to those followed by the Ottomans in the battle of Pelekanos. Describing one of the campaigns of Michael VIII in Asia Minor during the last years of his reign, Gregoras relates that, in 1280, when considerable Turkish military forces assembled in Paphlagonia, the emperor encountered them in the Sangarios river and decided to lead a large army against them. However, the Turks inflicted a crushing defeat on the Byzantines because the latter failed to react successfully to the battlefield tactics of their enemy. As Gregoras writes,

> They [the Turks] laid ambushes and waited for the Romans to arrive. They did not fight pitched battles but instead, as was their custom, they attacked in groups one after the other. They pretended to retreat. Then they returned with great speed. They did this repeatedly until they disrupted the order of their enemy and broke their ranks. Then, once their enemies were in disorder and disorganized, they charged and won very easily. They [the Turks] could not attack the Romans, who wore armour and had good weapons. They were fleeing and the Roman army was pursuing them unreasonably (it should not have done so) without good order, without suspecting anything and without any fear, despite the prohibitions and orders of the most prudent of the generals. But it seems that this war was the beginning of the disaster of the Romans, which was the divine punishment for countless sins. And when the Turks crossed Sangarios, they (the Romans) crossed it as well, killing those who could not ride fast, until suddenly and unexpectedly they fell in the ambushes laid by the Turks. And while they were gasping for breath, due to the long pursuit and the heat, the Romans had to face fresh and eager

[42] *Taktika*, XVIII. 43–72.

> enemies. Thus, they were encircled by countless enemy soldiers along the river and all apart from a few were slaughtered, achieving little that is worthy of mentioning and remembering.[43]

The mobility of the Turkish soldiers and their ability to successfully ambush the imperial armies is also reflected in the account of George Pachymeres. He reports that in 1301 a large number of Turks in the area between Nikomedia and Nicaea took by surprise an army led by Leo Mouzalon. As Pachymeres writes, when the Byzantines reorganised their forces, they pursued the Turks, who quickly withdrew to a higher point on a nearby mountain. From there, their archers inflicted heavy casualties on the Byzantine army. Eventually, the Byzantines were encircled and their general narrowly escaped with his life. Pachymeres concludes that this event paved the way for the further advance of the Turks, encouraged Osman to carry out more daring raids and motivated more Turkish warriors to join his forces. Pachymeres holds Leo Mouzalon responsible for the Byzantine defeat, blaming him for either failing to understand the battlefield tactics of the enemy, or for being unable to impose his authority and necessary discipline on his troops.[44]

The discussion above shows that the Byzantine military leadership was often unable to impose the necessary discipline on its troops and prevent them from falling into the ambushes of the Turkish troops. It is significant that, long before the thirteenth century and despite the recommendations of military treatises, the authors who saw as a cardinal sin the undisciplined rush to pursue an enemy fielding light cavalry archers described the general difficulties for Byzantine armies of dealing with opponents who made extensive use of archers. The author of the *Taktika* observes that Byzantine archery had been neglected in the period when the Arab forces recruited their substantial numbers of Turkish cavalry archers for use against the Byzantines. He comments that the decline of archery is one of the main reasons why the Byzantines had made so many mistakes on the battlefield and concludes that all Byzantines up to the age of forty should be forced to learn to wield the bow.[45] Indeed, there are contemporary examples showing the effectiveness of archery against the Byzantines. In the

---

[43] Gregoras, I, 139.

[44] Pachymeres, IV, 365.

[45] *Taktika*, VI. 5; W. Kaegi, "The Contribution of Archery to the Conquest of Anatolia," *Speculum* 39 (1964), 99–101.

eleventh century, the Byzantines failed repeatedly to deal effectively with the cavalry archers of the Seljuk armies.[46]

It has been argued that one of the main reasons that the Byzantines experienced great difficulties when confronted by effective archery during the middle Byzantine period is that this type of weapon was foreign to Byzantium, the tactical organisation of which did not require systematic practice of the required skills. Moreover, the cavalry archers employed by the Byzantines were mercenaries and allies. Consequently, the possibilities of this sort of training were limited.[47] With regard to the late Byzantine period, Asia Minor seems to have been a source of competent archers, most probably infantry, who seem to have made up a significant part of the armies of the so-called empire of Nicaea. Their presence is attested in the siege of Constantinople in 1235 by John III.[48] Describing the accession to the throne by Theodore II, Akropolites points out that Philadelphia's residents could bear arms. He adds that they were very competent archers and were in constant warfare with the Turks. He also comments that Theodore's army in his second campaign against the Bulgarians in 1256 included a great many archers and that the majority of the Byzantine soldiers who fought at the battle of Pelagonia in 1259 were archers.[49] Pachymeres relates that a great part of the forces of Michael VIII in the siege of Galata in 1260 were Nicaean archers, whose successful shots caused many problems to the besieged Latins. Referring to a rebellion against Michael VIII in Asia Minor, in 1262, Pachymeres comments that, 'the peasants in this area were competent archers.'[50]

However, there is no indication that the Byzantines employed an effective force of archers, either cavalry or infantry, against the Turks throughout the fourteenth century. It seems that during the first decades of the fourteenth century the role of Byzantine archers was reduced to one of garrisoning cities and fortresses. The loss of imperial control over Asia Minor, which was the main source of competent archers who were experienced in daily frontier warfare against the Turks, must have been one of the main reasons why their presence is not mentioned by

[46] See Kaegi "The contribution of Archery," 98–108.
[47] Haldon, *Warfare, State and Society*, 215–217.
[48] Da Canale Martino, *Cronaca Veneta*. G. Galvani (ed.), in *Archivio Storico Italiano* 8 (1845), 364.
[49] Akropolites, I, 105, 119, 169.
[50] Pachymeres, I, 173, 267.

the fourteenth-century sources. Nonetheless, this does not mean that archers, whether mounted or not, would necessarily have been effective in fighting the Turks. The troops of Choirovoskos, who in 1303 created an army and operated in Asia Minor, included many archers, but his operations failed.[51] Yet the Catalan Grand Company, which relied substantially on heavy cavalry and not on archery, achieved significant victories at the expense of the Turks.

Moreover, stereotyped views about the Turks affected the performance of the late Byzantine armies in the late thirteenth and early fourteenth centuries. It seems that the Byzantine leaders, even if they knew the fighting methods of the Turks, very often could not fully understand the strength of their armies and consequently underestimated them. For instance, Pachymeres, who otherwise displays a good knowledge of the military affairs of the Byzantine state and provides a reasoned account of the collapse of the Byzantine defence in Asia Minor, remarks that the Turks were fighting like brigands.[52] He also relates that Michael VIII underestimated the strength of their armies and believed that he could recover what had been lost in Paphlagonia whenever he liked.[53] In his description of the Turcoman emirates Metochites repeats stereotypical views about barbarians who live like beasts and in lawlessness.[54] This attitude was one of the reasons that led Michael VIII to transfer significant resources and manpower to the western parts of the empire, in order to organise the defence against a possible attack from Western European powers and to conduct military operations against the Epirots and the Latins in the Peloponnese. It seems that Michael VIII was more afraid of the Frankish cavalry of Charles I of Anjou than of the Turcoman cavalry archers. In other words, the feigned retreats and mobility of the nomads were seen by members of the Byzantine elite as acts of cowardice and brigandage, and not as the sort of military tactics that a well-organized army would follow. This shows that the Byzantine failure to react successfully to the battlefield tactics of the Turks could partly be described as a failure of cultural adaptation, reinforced by ethnic stereotypes about 'barbarians.'

The answer to the question why the Byzantines continuously made the same mistakes despite the instructions given by the military treatises

---

[51] Pachymeres, IV, 485–487.
[52] Pachymeres, I, 33.
[53] Pachymeres, I, 293.
[54] Metochites, *Orations*, 379–387; Metochites, *Miscellanea*, 729.

and despite numerous examples provided by the experience of fighting against enemies of similar nature in the past, is also likely related to the conservatism of the ruling elite, which monopolised the military commands. It would be difficult for the higher aristocracy, which had full control over the military organisation, to learn from experience. Therefore, the experience provided by examples, such as the difficulties the heavy Byzantine cavalry encountered in fighting the light Turkish cavalry in the eleventh century did not have any significant impact on the thought of the late Byzantine commanders, with the exception of experienced and highly educated generals such as Kantakouzenos, who, however, were unable to impose the necessary discipline on their troops. The military commanders were members of a separate social group, which tends to develop its own values and traditions.[55]

It seems useful to compare the difficulties the Byzantines had in encounters with the Turks with the warfare in the Middle East between the Crusader and Islamic armies. Due to the long distances between the population centres and the large tracks of desert both the Western and Islamic armies placed emphasis on mounted warfare. Although both armies were made up mainly of cavalry, the difference was that the Islamic armies used huge numbers of light cavalry and light archers. This resulted in the deployment of different tactics. While the crusaders sought to bring the battle to close quarters as soon as possible, the Islamic troops relied on ambushes and on firing from a distance. Like the Byzantines, quite often the Crusader armies broke their ranks, lost their discipline and were defeated.[56]

The difficulties European knights faced when confronted by Turkish light cavalry archers is reflected in the discussion among Frankish commanders, as is attested by the *Chronicle of the Morea*, on the eve of the battle of Prinitza (1262), where the Franks inflicted a heavy defeat on the Byzantines. According to the anonymous chronicle, the Frankish generals commented that the Byzantines would have never emerged victorious in Pelagonia 'if they lacked the arrows which slaughtered the horses.' To avoid repeating the same mistake, the Frankish generals decided not to attack the Byzantines with their lances. Instead, they decided to rely on the Turcoman mercenaries who had recently

[55] Haldon, *Praetorians*, 344.

[56] France, *Western Warfare*, 211–213; C. Hillebrand, *The Crusades. Islamic Perspectives* (Edinburgh, 1999), 511–514.

deserted the Byzantines and joined their side. Their aim was to shoot and kill the horses of the Byzantine army.[57] Therefore, the late Byzantine armies were not the only ones which had difficulties in dealing with battlefield tactics relying on the mobility of light cavalry and on the effectiveness of cavalry archers.

The employment of highly skilled and disciplined Latin heavy cavalry was the most effective measure the Byzantine government took to solve the problem of the tactical ineffectiveness against the Turks. Gregoras states that at the battle of Antioch-on-the Menander in 1211 the Byzantine troops took advantage of the fact that the Seljuk sultan avoided fighting in a narrow area because of the large number of his troops. Gregoras implies that this was a tactical mistake. Therefore, the battle took place on level ground, favourable for the use of heavy cavalry. The same historian describes how the 800 Latin mercenary heavy cavalrymen hired by Theodore I attacked and reached the rear of the Seljuk ranks, neutralising the enemy slingers and archers and taking advantage of their dense formation, not giving them the opportunity to react flexibly and use their bows.[58]

The Catalan Grand Company achieved significant success against the Turks. Gregoras comments that the Catalan company was a highly organized and very effective and disciplined military force.[59] Their high level of professionalism and experience, and the good co-ordination between cavalry and infantry and the maintenance of discipline, a field in which the late Byzantines constantly failed, brought successful results. As Muntaner states, in 1304, at Philadelphia the Catalan cavalry attacked the horse archers of the principalities of Saruhan and Aydın swiftly before the launch of their arrows could damage them. Once they brought the battle to close-quarters, the Almugavar infantry successfully supported the cavalry.[60]

The effectiveness of the use of disciplined Latin troops against the Turks is reflected not only in the success of the 'Italians' of Theodore I and of the Catalans, but also in the success of the armies of Amadeo of Savoy and Marshal Boucicaut (1366 and 1399). Amadeo VI of Savoy, a cousin of John V, leading around 1500 mercenary soldiers, conducted a private crusade, seized Gallipoli from the Ottomans, with the help

[57] *Chronicle of the Morea*, vv. 4920–4968.
[58] Gregoras, I, 19–20; Akropolites, I, 16.
[59] Gregoras, I, 221–222.
[60] Muntaner, II, 435–436.

of the Genoese ruler of Lesvos and returned it to Byzantium. He also recovered for Byzantium Mesemvria and Sozopolis, which he seized from the Bulgarians. Marshal Boucicaut was sent to Byzantium by the French king Charles VI in 1399. He was a famous soldier, with great experience in the east and he had fought in the battle of Nikopolis (1396). In 1399, he arrived in Constantinople leading 400 men-at-arms from, 400 armed attendants and a number of archers. His total force amounted to 1200 soldiers and was augmented along the way. In summer of 1399, Boucicaut and Manuel II launched a joint expedition on the fortresses of Black Sea close to Constantinople, which were held by the Ottomans and cleared Constantinople's hinterland. They also launched an unsuccessful attack on Nikomedia. Realising the ephemeral character of these successes, Boucicaut advised Manuel to write to France and ask for aid.[61]

However, the long-term employment and maintenance of large groups of high-quality Western European heavy cavalrymen to deal effectively with the Ottomans was something the Byzantines could not afford. In 1367, John V paid 15,000 *hyperpyra* to Amadeo as compensation for the wars he fought for Byzantium from money he had borrowed from the Genoese.[62] Temporary employment of very competent mercenaries could lead to successes, but the Byzantine state did not have the resources to turn temporary victories into permanent gains. Byzantium could not afford the long-term employment of expensive mercenaries of high quality.

Furthermore, the Byzantines did not use extensively mobile light cavalry units such as mounted crossbowmen and the *Tourkopouloi* (*Turcopoles* as they are called by the western sources).[63] The *Tourkopouloi* appear for the first time in Byzantium as part of the army of Alexios I Komnenos.[64] They were the offspring of a Byzantine mother and a Turkish father.[65]

---

[61] See E.L. Cox, *The Green Count of Savoy, Amadeus VI and Transalpine Savoy in the Fourteenth Century* (Princeton, 1967), 209–230; D. Nicol, *The Last Centuries of Byzantium* (London, 1972), 274–277; Barker, *Manuel II*, 62, 163.

[62] Cox, *Green Count*, 230.

[63] According to Marshall the *Tourkopouloi* fighting for the westerners in the east, were native Middle East troops or westerners using native military equipment. They were employed to counter Muslim's methods of combat and they also augmented the small numbers of the Latin armies: C. Marshall, *Warfare in the Latin East 1102–1291* (Cambridge, 1992), 58–60.

[64] *Gesta francorum et alliorum Hierosolimitanorum*, R. Hill (ed.), (London, 1962), 6, 9, 16.

[65] Hohlweg, *Beiträge*, 67–68.

The late Byzantine *Tourkopouloi* were the Christianised Turks, who had initially arrived in Constantinople in the retinue of the Seljuk Sultan Izz al-Din and some of them had decided to stay in Constantinople after the Sultan's flight in 1264. For the last time they appear in 1305 battle of Apros where they deserted the Byzantines.[66] Villehardouin states that the Latins of Constantinople employed *Tourkopouloi* who came from Syria to reinforce the army of the Latin empire. He adds that in 1207 *Tourkopouloi* and mounted crossbowmen were sent to scout around Stenimachos in Thrace.[67] There is not any evidence that any of these soldiers were recruited by the Nicaean rulers.

The aforementioned battle of Pelekanos is the last attested large-scale encounter between the Byzantines and the Turkish cavalry archers in Asia Minor. After this battle, Kantakouzenos' and Gregoras' accounts mention numerous Turkish raids on Thrace and Macedonia. Usually, it is not specified whether these raiding parties were Ottoman or not. It is noticeable that, quite often the Byzantine sources mention the presence of large numbers of foot soldiers in the Turkish raiding armies. For instance, the army of Umur, the emir of Aydın, which fought on Kantakouzenos' side in the civil war of 1341–1347, was made up mainly of foot soldiers.[68] Moreover, by 1400 major administrative developments had been completed in the Ottoman state by the establishment of the *timar* cavalry and *janissary* troops. The *Timar* created the obligation of its holder to serve wherever required, while the *janissaries* were a standing Ottoman army armed by the central government and one of their main tasks was the garrisoning of fortresses.[69]

The Byzantine government after the middle fourteenth century could not afford similar developments. The reduction of the territory under imperial control resulted in the drastic shrinking of the necessary resources for the maintenance of an effective military force. Training and armament remained a privilege for the aristocracy and there was not any attempt to increase the army by systematically

---

[66] Pachymeres, IV, 573, 601, 627; Bartusis, *The Late Byzantine Army*, 62.

[67] Villehardoun, 96, 188; J. Longnon, *L' empire Latin de Constantinople et la principauté de Morée* (Paris, 1949), 67; B. Hendrickx, *Οι πολιτικοί και στρατιωτικοί θεσμοί της Λατινικής Αυτοκρατορίας της Κωνσταντινουπόλεως κατά τους πρώτους χρόνους της υπάρξεως της* [*The Political and Military Institutions of the Latin Empire of Constantinople during the Early Years of its Existence*] (Thessalonica, 1970), 153.

[68] Kantakouzenos, II, 392.

[69] Imber, *The Ottoman Empire*, 256–263; Kafadar, *Between Two Worlds*, 17; Lowry, *Ottoman Balkans*, 3–9.

recruiting and training troops from the lower social classes. The Ottoman technological and administrative advantage in the fifteenth century is reflected in the account of Laonikos Chalkokondyles. He provides information on the training, recruitment, finance and role of the various types of troops of the Ottoman army. Chalkokondyles points out that semi-independent frontier lords, contributed decisively to the expansion of the Ottomans in the Balkans. They had under their command the so-called *akıncı* raiders. The *akıncı* and their leaders spearheaded the Ottoman expansion in the Balkans and created the infrastructure which ensured the permanent nature of the Ottoman conquest.[70] Chalkokondyles calls them *hippodromoi* (fast horsemen) and remarks that they were maintained exclusively through the booty they captured.[71] Manuel II, who was an eyewitness of the organisation and tactics of the Ottoman armies, stresses the quality of training of the Ottoman troops, stating that they could bear hardships for a long period of time and could remain in a hostile land far longer than other armies who would not have remained even in a territory which produces every kind of goods.[72]

The development and superiority of the Ottoman army is reflected in the lack of effective resistance against the Ottoman raids in the Peloponnese in the fifteenth century. The inability of the Moreots to resist the Turks effectively was the result of the lack of strong centralised authority that would force the local lords to contribute to the defence of the Morea and of the technological superiority of the Ottomans who also had at their disposal far superior resources compared with the Byzantines. It is interesting that Chalkokondyles' account of the raid Murad II led in 1446 in the Peloponnese focuses on the large amount of supplies of the Ottoman forces. He mentions the presence of a large market in the Ottoman camp where plenty of food supplies, fodder and horses were available and remarks that many of the powerful men who followed the Sultan on his campaigns were leading their own baggage trains. Similarly, Chalkokondyles points out the huge number of pack animals that were used to supply the Ottoman army in the siege of Constantinople in 1453.[73]

---

70 Lowry, *Ottoman Balkans*, 10.

71 Chalkokondyles, I, 93; Doukas (175) gives a detailed description of the organisation of a Turkish raid, the *akın*.

72 Manuel II, *Funeral Oration*, 160.

73 Chalkokondyles, II, 149–150.

Infantry

A major constraint on the study of the role of infantry troops in late Byzantine warfare are the prejudices of the sources towards foot soldiers, who were of lower social status than the cavalry troops.[74] Nevertheless, the sources do not omit to mention the presence of foot soldiers on the battlefield, but their role is portrayed as unimportant for the outcome of battles. Similarly, the sources never state the names and the deeds of individual foot soldiers. Moreover, the Byzantine aristocracy was conscious of its status as a social class. Therefore, it is logical to assume that the social gap between the wealthy cavalrymen and the infantry soldiers in late Byzantium widened. Kantakouzenos, an aristocrat soldier himself, does not omit to mention the presence of infantry forces, although throughout his account he emphasises the social differentiation between cavalry and infantry troops. When he states that he planned to campaign in Tzepaina and Stenimachos in 1343, he points out that the military forces of this area consisted of no fewer than 1000 cavalry and a large number of infantry. Shortly afterwards he comments that Tzepaina and Stenimachos before the war had more than 1500 good soldiers (*stratiotai*). These were cavalrymen and *pronoia* holders.[75] He ignores the infantry troops he had previously mentioned. The *Chronicle of the Morea*, being closer to Western European ideals of warfare ignores the role of the infantry troops, giving the impression that the outcome of the military operations in the Peloponnese was determined exclusively by cavalry soldiers, particularly by the bravery of individual knights.

Usually, the foot soldiers are portrayed as ineffective and low quality troops, sometimes hastily mustered. The *Chronicle of the Morea* mentions that in 1262, Makrenos in the Peloponnese had under his command massive infantry forces, which he recruited locally.[76] However, when it comes to the description of battles and tactics, the *Chronicle of the Morea* gives the impression that everything depended on cavalry formations and tactics. Nonetheless, oftentimes, when the sources are critical of the quality of the infantry troops, they reflect real problems

---

[74] For a discussion of the effectiveness and use of the infantry between the fifth and twelfth centuries, with analysis of the military treatises see Haldon, *Warfare State and Society*, 190–214.

[75] Kantakouzenos, II, 405, 406.

[76] *Chronicle of the Morea*, vv. 4655–4663.

rather than biased views. Gregoras states that in1273, on his way to the campaign in Thessaly John Palaiologos recruited infantrymen from Thrace and Macedonia. They do not seem to have been high quality troops and their morale was low, since they easily lost their discipline and coherence once they were attacked by the Frankish cavalry of the ruler of Athens Jean de la Roche (1263–1280).[77] Possibly, they did not even fight, since according to Pachymeres, once the first rank (he does not specify who they were) was defeated, the rest of the army immediately fled.[78] In 1305, the foot soldiers in the battle of Apros made little impact on the outcome of the battle and fled once the *Tourkopouloi* and the Alans deserted the Byzantines.[79] In 1311, Michael IX was sent to fight the Turcoman adventurer Halil, leading a force of non-professional inexperienced infantry troops. Gregoras specifies that they were not soldiers but rather 'a rural mob interested only in looting.'[80]

Nevertheless, the role of the foot soldiers was crucial when they were placed under the command of effective leadership and when they were used properly and in correspondence with the geographical and tactical demands of the battlefield. Gregoras praises the role of the infantry troops in the victory of Philes Palaiologos against the Turks of Halil. He points to the contribution of Philes to the morale of his troops, who encouraged the infantry soldiers to attack first and fight gallantly 'in a very bloody and long battle.'[81] Infantry, especially archers, when properly used could achieve significant victories against heavy cavalry formations. Such was the case in the battle of Pelagonia in 1259. Akropolites states that the Byzantine army included a considerable number of Byzantine archers, who alongside the Cuman and Turkish cavalry archers repeatedly ambushed the enemy troops, prohibiting them from reaching water supplies and attacking their baggage train.[82]

---

[77] Gregoras, I, 111; Pachymeres (II, 421–431) does not mention the presence of infantry troops.

[78] Pachymeres, II, 425.

[79] Pachymeres, IV, 599; Gregoras (I, 230) mentions only the presence of infantry troops not their role, while Pachymeres does not make any reference to the presence of infantry. The cross-examination of the information provided by Gregoras and Pachymeres leads to the conclusion that probably the Vlachs, the *Thelematarioi*, and the refugees from Asia Minor fought on foot.

[80] Gregoras, I, 256.

[81] Gregoras, I, 267.

[82] Akropolites, I, 169.

Foot soldiers were very useful when the geographical circumstances and the nature of the enemy did not allow the deployment of cavalry forces. In 1255, Theodore II deployed infantry tactics to deal with the Bulgarians in Rupel in Macedonia. Akropolites states that Rupel was protected by the Strymon and by two mountains, and that the Bulgarians had closed the very narrow pass. Theodore II sent part of the infantry forces to attack the right wing of the Bulgarians from the mountain, from a higher point. Akropolites states that the mountain was full of trees and suited to infantry operations. The cavalry, under the command of the emperor, joined the battle at a later stage.[83] In a further example demonstrating the usefulness of foot soldiers, in 1338 to suppress the rebellion and raids of the Albanians in Bellagrada and Kanina the Byzantines decided to deploy Tukish light infantry and archers sent by Umur. Kantakouzenos states that the Albanians were settled in mountains and had many places of refuge; it was very difficult for the cavalry to engage them. The situation was more difficult in the summer, when the Albanians were able to reach the highest points of the mountains, where even the infantry would find it difficult to advance, due to the natural fortifications, since 'the defenders were standing above the heads of the attackers.'[84] Kantakouzenos' description of the heavy defeat of the Albanians should not lead to the conclusion that there were no Byzantine infantry troops available. The Byzantine leaders decided that the Turkish infantry was more suited for this operation than the Byzantine foot soldiers. Moreover, since these troops were funded by Umur and supported by the reward of a large amount of booty, it would have been more expensive for the Byzantine state to employ native troops and to maintain their logistical support. Another example indicating the effectiveness of infantry troops in battlefield tactics was the victory of Matthew Kantakouzenos against Turkish raiding forces in 1348. Matthew followed the Turks through mountainous paths, planning to surprise them. This geographical background was favourable for the deployment of infantry. Gregoras remarks that Matthew had placed them on the sides of the ranks, while the higher quality cavalry troops were placed in the centre. Despite Gregoras' statement, that the cavalry was of better quality, the infantry made many attempts to break through the ranks of the Turks

[83] Akropolites, I, 116.
[84] Kantakouzenos, I, 496.

and paved the way for a cavalry attack.[85] The examples cited indicate that late Byzantine generals possessed the necessary knowledge to use their troops according to the requirements of the geography of the military operations they undertook.

The presence of infantry troops seems to have been increased in the course of the fourteenth century. Kantakouzenos' account shows that, a great part of the troops involved in the military operations after 1341 were foot soldiers and indicates that the infantry had started playing an increasingly significant role in the Byzantine military operations. The main reason for this development is that warfare in the period involved many large and small-scale sieges and raids rather than pitched battles. Kantakouzenos' account does not indicate any change in the military administration, tactics, and cultural perceptions about warfare, which would imply more extensive and sophisticated deployment of infantry tactics, parallel to contemporary developments in Western Europe. It was rather the lack of resources, both in terms of manpower and finance, as well as the nature of the military operations that led to the increase of the importance of infantry troops, who quite often were not trained soldiers but individuals, who hoped to take advantage of the abnormalities caused by the civil wars. For instance, the infantry soldiers of the brigand Momčilo, who were employed by both sides in the civil war of 1341–1347, were irregular troops.[86] The foot soldiers recruited by Kantakouzenos for the defence of Constantinople against the Genoese in 1348 were not regular and trained troops, while the information on the identity and quality of the troops led by Manuel Asan in 1350 in the conflict with the Genoese of Pera and of the infantry John V had under his command in the conflict against Matthew Kantakouzenos in 1355, is insufficient to come to a conclusion.[87] As for results, Kantakouzenos' troops in 1348 proved inexperienced. Asan's were not tested on the battlefield, since they did not fight, and John V avoided any engagement for tactical reasons. He did not want to deploy his infantry on flat terrain.

The mid-fourteenth century saw the increasing significance of the infantry troops in Western Europe. This was partly the result of the development of centralised states.[88] Furthermore, in the course of

[85] Gregoras, II, 836; Kantakouzenos, III, 66.
[86] Kantakouzenos, II, 403.
[87] Kantakouzenos, III, 39, 196, 320.
[88] France, *Western Warfare*, 132.

the fourteenth century, the outcome of a large number of battles in Europe enhanced the role of the infantry.[89] In late Byzantium there are no references to mass infantry contingents, since the late Byzantine state did not establish any form of universal or territorial obligation to military service which would yield foot soldiers.[90] Nor was the state able to afford the logistical needs of these troops, even though they were fewer than the needs of the cavalry soldiers. The only exception was Thomas Magistros' suggestion that all the citizens should be trained in arms.[91] However, he is influenced by the specific circumstances of the conflict between Byzantium and the Catalans. The defence of the Byzantines against the Catalans in Macedonia relied on the strength of their fortifications and not on successes on the battlefield. In addition, infantry soldiers could be effective only when recruited in large numbers and were able to maintain high levels of discipline on the battlefield. This can be achieved through intensive training, which means that the throne possessed the necessary resources to recruit and maintain a large number of men.[92] This was not the case with late Byzantium, which although was a centralised state, it did not have the financial capacity to maintain such troops andthe continuous territorial reduction resulted in the decrease of the available manpower.

---

[89] See K. De Vries, *Infantry Warfare in the Early Fourteenth Century* (New York, 1996), passim.

[90] For the presence of massive infantry troops in Western Europe see France, *Western Warfare*, 131–133.

[91] *PG* 145, col. 509.

[92] P. Morillo, "The 'Age of Cavalry' Revisited," in D.J. Karay – L.J.A. Villalon (eds.), *The Circle of War in the Middle Ages. Essays on Medieval Military and Naval History* (Woodbridge, 1999), 52.

CONCLUSION

In the Balkans, Frankish Greece and Asia Minor, most of the period between 1204 and 1453 witnessed political fragmentation, insecurity and ceaseless warfare between small political entities, which had too few resources to control large territories, conduct long campaigns and recruit large armies. Byzantium was part of this fragmented world, above all after the loss of Asia Minor in the first decades of the fourteenth century.

Despite the transformation of Byzantium into a second- or third-rate power and the evolution of Constantinople into a city-state, there was little change in the traditional Byzantine ideological approaches towards warfare. The imperial prerogative of waging war and constructing fortifications was never doubted, although political and military realities compromised it, allowing local and private authorities to act independently, building and instigating the defence of fortresses and towers. The traditional idea that Byzantium fought Just Wars against enemy assaults, invariably for the recovery of imperial land, features prominently in the political and ideological agenda of the late Byzantine rulers. Notwithstanding this continuity, analysis of the available source material shows that Byzantine ideas about warfare varied in the reigns of successive emperors and in accordance with the political and military circumstances and priorities of the throne. Moreover, like their predecessors, the later Byzantines did not fully develop a notion of Holy War. Nevertheless, religion was used as a means of diplomacy; for example, to avert any possible Latin attack for the recovery of Constantinople and to attract Western military support against the advancing Ottomans. However, given the polyglot and multi-ethnic character of the late Byzantine army, as well as the fact that it was composed mainly of full-time mercenary and *pronoia* holders, it is highly unlikely that the official justifications of military conflict had much impact on the individual soldiers. Moreover, it is hard to know the motives of each individual soldier and most of the historians of the period, including the clerics, point to material profit, in the form of grants of land, *pronoiai*, higher salaries, booty and promotion to higher office as most important in the motivation of soldiers.

With extremely few exceptions, members of the higher aristocracy monopolised the military command. Military power enabled late Byzantine aristocrats to increase their influence over the ruler and improve their position in the court hierarchy. The military character of late Byzantine aristocracy reinforced the development of an ethos which promoted heroic individualism and valour in combat. This is reflected in the surviving poems of Manuel Philes, in the late Byzantine romances and in narrative histories, the authors of which describe the heroic acts of individual aristocratic commanders. Byzantine military aristocrats and rulers, such as Kantakouzenos and Andronikos III, were attracted to chivalric ideas as they developed in the West and were willing to adopt practices that promoted military prowess, such as tournaments and jousting. Moreover, late Byzantine aristocrats were familiar with the significance and use in Western Europe of heraldic emblems. Unfortunately, the extent to which such emblems were adopted by the Byzantines remains unknown.

The soldiers who were employed by the late Byzantine state can be divided into two broad categories: soldiers who were maintained from incomes derived from the exploitation of land and mercenaries. However, a rigid typology of soldiers is impossible. For instance, there were mercenaries who received grants of land and groups of soldiers whose funding remains unknown. It seems that the soldiers who were maintained through *pronoia* grants were the most effective part, though not the majority, of the native soldiery. However, the lack of sources, combined with the inconsistent and archaic terminology used by contemporary authors, prevents us from reaching safe conclusions concerning their organisation and role in the military operation of the Byzantine armies. The limited available information suggests that the task of the soldiers who were maintained through *pronoiai* was to participate in campaigns and to support the local defence of the province or city where they were established. The continuous conflicts between aristocratic cliques, the steady territorial reduction of the Byzantine state and the advance of its enemies meant that very often the state was unable to defend the estates and interests of its soldiers. As a result, in times of crisis many of these soldiers supported rebellious generals, while others joined the expanding enemies of Byzantium, or abandoned military service all together.

Unlike the pronoiars, mercenaries have been blamed for being disloyal to the throne and for contributing to the final collapse of Byzantium. It has been argued that after 1204 the army had been transformed into agglomerates of foreign mercenaries seeking temporary employment.[1] However, mercenaries provided the army with troops constantly ready to campaign. Moreover, they covered the tactical insufficiencies of the late Byzantine army. On many occasions, the Byzantine government hired highly skilled specialised professional warriors and troops of high quality, who proved effective on the battlefield. In addition, the employment of mercenaries in the late Byzantine period should be examined in the context of the increased military professionalism at the time in Western Europe, where technological developments in the art of war encouraged specialisation and professionalism. However, in some cases, in the early fourteenth century in particular, the state made a series of strategic and administrative mistakes, such as overestimating its power to pay for expensive, though high-quality, mercenaries and failing to keep balanced relations either between the large groups of mercenaries or between the foreign mercenaries and the native troops. The negative image of the mercenaries, for example, was coloured by the friction between the government, the Catalan Grand Company and the Alans. However, the need for mercenaries, as well as their value and usefulness, was never doubted. Moreover, the negative comments about groups of mercenaries, made by authors critical of the financial policies of particular emperors or influenced by classical models, should not lead to the conclusion that the presence of mercenaries was detrimental to the affairs of Byzantium.

Examination of the military campaigns conducted by the later Byzantines reveals a simplification of the administrative infrastructure and the limited geographical scope of the expeditions undertaken by the late Byzantine armies. Like all small political entities in the Balkans and Asia Minor, Byzantium used small armies for these operations, often recruited along the way. It seems that in the fourteenth and the fifteenth centuries these armies were supported by small baggage trains and it was expected that most of the soldiers would cover their expenses through their own means. The imposition of the *mitaton* on the local population and the use of markets, together with the capture and distribution of spoils of war, covered a large part of the maintenance of campaigning forces.

---

[1] S. Vryonis, "Byzantine and Turkish Societies and their Sources of Manpower," in V. Parry (ed.), *War Technology and Society in the Middle East* (London, 1975), 128.

The endless military conflicts between numerous small and poor states increased the importance of fortifications as a means of defence. As a result, sieges and blockades of fortresses, citadels, and fortified cities and towers outnumbered pitched battles. Most of the sieges described by the sources were short operations, which involved small armies of rarely more than a few hundred soldiers. The aim of the besiegers was to undermine the economic basis and morale of the defenders by ravaging their countryside and destroying the agricultural hinterland of the besieged city. This would compel the city to surrender without any fighting, thus saving the attackers from the expense of fighting a pitched battle that might cost much in resources and manpower.

The sources mention the use of siege engines by the Byzantines and their enemies. However, the archaic and inconsistent terminology makes it impossible to assess their impact and destructive power with confidence. What the sources make clear is that both besiegers and besieged used a small number of siege engines, which more often than not did not determine the outcome of sieges. Instead, most of the known successful operations of this kind were the result of negotiations and agreements. Gunpowder artillery started to make an impact on the military conflicts of the area in the first decades of the fifteenth century. Its use was rather expensive for a small and impoverished state such as fifteenth-century Byzantium. Nor did gunpowder artillery correspond to the nature of warfare or the strategic aims of the small political entities in Frankish Greece and the southern Balkans. The limited available evidence shows that the Byzantines adopted gunpowder artillery in the early 1420s or a little earlier. By 1453 they were aware of the advantages of the new technology. However, the accounts of the final fall of Constantinople indicate that the Byzantines had still not acquired a comprehensive knowledge of the deployment of canons.

The surviving descriptions of battles show that the Byzantines, like their predecessors, tried to exploit the tactical weaknesses of their enemies. Being aware of the development and effectiveness of heavy Western European cavalrymen, the Byzantines avoided fighting them on the open field and in pitched battles. Instead, they adopted the tactics of guerilla warfare, aiming at disrupting the order and breaking the discipline of their Western European opponents. To achieve this, the Byzantines used a significant number of light cavalry archers, mainly Turkish and Cuman mercenaries, whose tactics depended on ambush and feigned retreats aiming at the disruption of the enemy's order and at exhausting the heavily armoured Latin cavalry. The tactical aim of the Byzantines against the light cavalrymen of the East was to draw the enemy into close action as swiftly as possible. Byzantine military leadership was often unable to impose the necessary discipline on its troops and prevent them from falling into the ambushes of the enemy's light cavalry archers. This shows that the outcome of a battle depended heavily not on the soundness and knowledge of battlefield tactics, but on the ability, skills and training of soldiers to carry them out.

Infantry soldiers were very useful when they were placed under the command of effective leaders and when they were used properly and in response to the geographical and tactical demands of the battlefield. However, a significant late medieval development which the Byzantines failed to follow was the increasing significance of infantry troops in Western Europe. Since the middle of the thirteenth century the presence of massive infantry contingents, deployed in more sophisticated ways, had been a European-wide phenomenon. This was partly the result of the development of centralised states, which were able to gather large armies from a wider recruiting base, producing increased numbers of foot soldiers.[2] Byzantium continued to be ruled by a strong central government. Even during this period the economic powers of the Byzantine state, in terms of the collection and redistribution of tax resources, were immense. However, the taxable territory was small, fragmented and was constantly shrinking, while the reigns of the state often fell into the hands of aristocratic cliques. Consequently, it was impossible for geographical reasons for the state to assemble a large force, and the central government could not exact sufficient resources to muster a large army. The late Byzantine state had too few resources in terms of finance and manpower to recruit and train large numbers of infantry soldiers. Moreover, the conservatism of the social elite, which monopolised the military commands, should also be borne in mind. Kantakouzenos, who is our main source for the study of fourteenth-century warfare in the wider Byzantine world, gives the impression that it was inconceivable for a proper Byzantine soldier to fight on foot. In addition, the idea of universal military service was unknown in Byzantium.

Very often, the government is blamed by the authors of the period for mismanaging resources and for maladministration. For instance,

---

[2] France, *Western Warfare*, 132.

governors and military leaders of the Asian provinces during the reign of Michael VIII are blamed for alienating a significant part of the local military forces by keeping the lion's share of the booty for themselves. Pachymeres argues that this made many local soldiers join the enemy. Given the importance of booty in supplying and motivating late Byzantine armies, this was a serious mistake. Michael VIII is also held responsible for alienating the population of Asia Minor by imposing heavy taxes and denuding the eastern defences by transferring significant manpower from Asia Minor to Europe and thus facilitating the Turkish advance. Like his father, Andronikos II is also blamed for mismanaging the available resources. One of his most serious mistakes seems to have been the scaling down of the fleet in 1285. Supposedly, the government hoped by this action to save enough money to fund the defences of Asia Minor. However, it effectively increased Byzantium's dependence on foreign fleets. It encouraged the Turcoman emirates to develop naval forces, which enabled them to carry out extensive and damaging raids on Byzantine possessions. Andronikos II is also blamed for overestimating the financial strength of the empire and hiring rather expensive large groups of self-interested foreign mercenaries over whom the state could exert very little control, while the native soldiery were neglected. The decline of the native army from the 1290s to the 1310s reflects a conflict between the military elite and the government of Andronikos II. This conflict led to the imprisonment and disgrace of the most prominent generals, including the emperor's close relatives, and to rebellions. The crisis contributed to the weakening of the Byzantine defences in Asia Minor, provoking the Turkish advance. It would be superfluous to repeat that the civil wars of the fourteenth century exhausted the state's resources, which had fallen into the hands of aristocratic cliques. Therefore, the limited resources available to the state, maladministration, the mismanagement of resources and internal conflicts prevented the empire from forming a coherent and long-term military policy and from resisting its enemies effectively. Kantakouzenos writes that, by the time he abdicated in 1354, the state had no income and it was impossible to face the Turks on the battlefield.

It has been argued that a basic characteristic of the army during the thirteenth and fourteenth centuries was a growing numerical weakness.[3]

[3] L.-P. Raybaud, *Le gouvernment et l' administration centrale de l' empire byzantine sous les premières Paléologues* (Paris, 1968), 243.

Indeed, the late Byzantine state recruited considerably smaller armies than the eleventh and twelfth-century Byzantine empire. However, as Bartusis noted, the numerical weakness of an army is linked to the strength of opposing armies and to the political environment.[4] The political and military context during the period under discussion was completely different from that of the eleventh and twelfth centuries. In terms of size and strength the armies of the states of Epiros and Thessaly, the principalities of the Toccos and the Acciauiolis, the Latin principality of Achaia, the Genoese of Pera and the Bulgarians were not superior to those of Byzantium. In purely military terms, the late Byzantine state lacked neither creative military and strategic thinking nor military leadership capable of boosting the morale of its soldiers and inspiring them. Byzantine generals such as John Palaiologos, John Kantakouzenos and Manuel II seem to have possessed sound tactical knowledge and principles outlined by the military treatises of earlier epochs are echoed in their writings. That lineage was the decisive factor in the appointment of military commanders, not military ability and experience, should not be seen as a disadvantage, since this was the case with all contemporary armies. The late Byzantine armies were not always defeated. However, the size of a victory is reckoned according to the relatively large or small proportion of the enemy forces destroyed or captured and the importance of the economic resources captured. In this respect, the Byzantines consistently failed to inflict a decisive defeat on their enemies. What the late Byzantine armies lacked were sufficient resources, which would enable them to respond efficiently and swiftly to any challenge. It may be useful to repeat here some examples, which have been mentioned above in a different context. John Palaiologos' operations and presence in the area around the Meander in the 1260s were successful, but once he was recalled and sent to Europe, the Turks recovered the territory that the Byzantines had seized. Manuel II's successes in 1383 could not have any long-term effect, since the Ottomans had the resources to send a stronger force to defeat him. It seems that proportionately the Byzantines, even when victorious on the battlefield, spent more resources than some of their defeated enemies, such as the armies of the Turcoman principalities.

[4] Bartusis, *The Late Byzantine Army*, 347.

GLOSSARY

Archon	a term used for any official who possessed power.
chrysobull	A generic name for several types of documents bearing the emperors gold seal (*chryse voula*).
despot	a title created in the twelfth century. It was held by sons and brothers of the emperors; or a designation for the rulers of semi-independent territories, such as Epiros and the Morea.
kephale	literary meaning 'head'. From the middle of the thirteenth century the term *kephale* denoted the highest functionary of provincial administration in charge of the civil and military administration of a province or city. The office of *kephale* replaced that of *doux*.
megas domestikos	the commander-in-chief of the Byzantine army after the emperor.
megas logothetes	since the reign of Alexios I Komnenos the highest office in the civil administration of the Byzantine empire.
megas stratopedarches	a high-ranking military officer in charge of the supply system of the Byzantine army.
mesazon	the emperor's confidant entrusted with the administration of the empire.
nomisma, hyperpyron	nomisma meaning 'coin' in the earlier periods was used to denote the standard 24 keratia coin. *hyperpyron* was a gold coin. Both terms continued to be used although from the middle of the fourteenth century the empire ceased to struck gold coins.
paroikoi	The general name for the dependent peasants. In the later Byzantine period almost all peasants appear to have been *paroikoi.*
pinkernes	'cup bearer.' Initially a palace eunuch. The importance of the office grew significantly in the late Byzantine period and became a high honorific title.
porphyrogennetos	an imperial epithet meaning 'purple born' and designating a son or a daughter born after the father had become emperor.

praktiko a document drawn up by fiscal officials listing obligations of tenants on estates

prostagma a term designating an administrative order. Usually a short imperial document.

protosevastos a honorary title.

theme Initially a term used for a military division and for a territorial unit. In the Palaiologan period it was used for fiscal units.

BIBLIOGRAPHY

Primary Sources

Actes de Chilandar, M. Živojinović - V. Kravari - C. Giros (eds.), (Paris, 1998).
Actes de Docheiariou, N. Oikonomides (ed.), (Paris, 1984).
Actes de Xénophon, D. Papachrysanthou (ed.), (Paris, 1986).
Actes de Zographou I. Actes Grecs, W. Regel - E. Kurtz. - B. Korablev (eds.), (St. Petersburg, 1907).
Akropolites, George, *Opera*, 2 vols. A. Heisenberg (ed.), (Leipzig, 1903, repr. Stuttgart, 1978).
Alexander, P.-J. "A *chrysobull* of the Emperor Andronikos II Palaiologos in Favor of the See of Kanina in Albania," *B* 15 (1949–1951), 167–207.
Aşık Paşazade, *Osmanoğulları' nın tarihi*, K. Yavuz - M.A.Y. Saraç (eds.), (Istanbul, 2003).
Athanasios I, Patriarch, *The Correspondence of Athanasios I Patriarch of Constantinople. Letters to the emperor Andronikos II, Members of the Imperial Family and Officials*, A.-M. Talbot (ed.), (Washington, 1975).
Attaleiates, Michael, *Miguel Ataliates Historia*, I. Pérez Martin (ed.), (Madrid, 2002).
Bakker, W. - van Gemert, A. (ed.), *Ιστορία του Βελισσαρίου* [*The Story of Belissarios*] (Athens, 1988).
Blemmydes, Nikephoros, *Oratio qualem oporteat esse regem*, *PG* 142, cols. 611–658.
——, *Autobiographia sine curriculum vitae*, J. Munitiz (ed.), (Leuven, 1984).
Boissonade, J.F. *Anecdota Graeca*, 6 vols. (Paris, 1829–1844 repr. Amsterdam, 1962).
Brocquière, Bertrandon de la, *La voyage d' Outremer de Bertrandon de la Broquière*, Ch. Schefer (ed.), (Paris, 1892).
Burgman L. - Magdalino P. "Michael VIII on Maladministration. An Unpublished Novel of the Early Palaiologan Period," *FM* 6 (1984), 377–390.
Chalkokondyles, Laonikos, *Laonici Chalcocandylae Historiarum demonstrationes*, 2 vols. E. Darkó (ed.), (Budapest, 1922).
Choniates, Niketas, *Nicetae Choniatae Orationes et epistulae*, J.L. van Dieten (ed.), (Berlin, 1973).
——, *Nicetae Choniatae Historia*, J.L. van Dieten (ed.), (Berlin, 1975).
Chronicle of the Morea, J. Schmitt (ed.), (London, 1904); tr. Luriel, H. *Crusaders as Conquerors* (London, 1964).
Chrysostomides, J. (ed.), *Monumenta Peloponnesiaca* (Camberley, 1995).
Clari, Robert de, *La conquête de Constantinople*, P. Lauer (ed.), (Paris, 1974).
Clavijo, Ruy González de, *Embassy to Tamerlane*, tr. Guy Le Strange (London, 1928).
Constantine VII, Porphyrogennetos, *De Ceremoniis aulae byzantinae*, 2 vols. I. Reiske (ed.), (Bonn 1829–30).
——, *Three Treatises on Imperial Military Expeditions*, J. Haldon (ed.), (Vienna, 1990).
Constantinides-Hero, A. *The Life and Letters of Theoleptos of Philadelphia*, (New York, 1994).
Cronaca dei Tocco di Cefalonia di Anonimo, G. Schiró (ed.), (Rome, 1975).
Da Canale, Martino, *Cronaca Veneta*, G. Galvani (ed.), *Archivio Storico Italiano* 8 (1845), 268–766.
Dagron, G. - Mihaescu, H. (eds.), *Le traité sur la Guérilla (De velitatione) de l' empereur Nicéphore Phocas (963–969)* (Paris, 1986).

Dennis, G.T. *Three Byzantine Military Treatises* (Washington, 1985).

Directorium ad faciendum passagium transmarinum, C.R. Beazley (ed.), *AHR* 13 (1907–1908), 66–115.

Doukas, *Istoria Turco-Bizantina (1341–1462)*, V. Grecu (ed.), (Bucharest, 1958).

Eustathios of Thessalonica, *La espugnazione di Tessalonica*, S. Kyriakidis (ed.), (Palermo, 1961).

Gautier, P. (ed.), "Un récit inédit sur la siege de Constantinople par les Turcs (1394–1402)," *REB* 23 (1965), 100–117.

Gesta francorum et alliorum Hierosolimitanorum, R. Hill (ed.), (London, 1962).

Gregoras, Nikephoros, *Nicephori Gregorae Byzantina Historia*, 2 vols. L. Schopen (ed.), (Bonn, 1829–1855).

——, *Logos prosphonematikos*, A. Westermann (ed.), *Excerptorum ex Bibliothecae Pauline Lipsiensis libris manuscriptis pars prima* (Leipzig, 1864).

——, *Nicephori Gregorae epistulae*, 2 vols. A. Leone (ed.), (Lecce, 1982).

Guillou, A. (ed.), *Les archives de Saint-Hean Prodrome sur le mont Ménécée* (Paris, 1955).

Heisenberg, A. *Quellen und Studien zur spätbyzantinischen Geschichte* (London, 1973).

Holobolos, Manuel, *Manuelis Holoboli orationes*, M. Treu (ed.), (Potsdam, 1906–1907).

Joseph the Philosopher, *Summation of Rhetoric*, C. Waltz (ed.), *Rhetores Graeci*, 9 vols. (Stutgard/Tubingen, 1832–1836).

Kananos, John, *De Constantinopolis obsidione*, E. Pinto (ed.), (Napoli, 1968).

Kantakouzenos, John, *Ioannes Cantacuzeni eximperatoris historiarum libri IV*, 3 vols. L. Schopen - B. Niehbur (eds.), (Bonn, 1828–1832).

Kekaumenos, *Cecaumenos strategikon*, B. Wassiliewsky - V. Jernstedt (eds.), (St. Peterburg, 1896).

Komnene, Anna, *Alexias*, D.R. Reinsch - A. Kambylis (eds.), (Berlin, 2001).

Kresten, O. *et als.* (ed.), *Das Register des Patriarchats von Konstantinopel 1315–1355*, 3 vols. (Vienna, 1981–2001).

Kriaras, E. (ed.), *Τα μεσαιωνικά ιπποτικά μυθιστορήματα* [*The Medieval Chivalric Romances*] (Athens, 1955).

Kritoboulos, *Critobuli imbriotai historiae*, P. Reinsch (ed.), (Berlin, 1983).

Kydones, Demetrios, *Démétrius Cydonès correspondance*, 2 vols. R.-J. Loenertz (ed.), (Vatican, 1956, 1960); tr. F. Tinnefeld, *Demetrios Kydones Briefe*, 4 vols. (Stuttgard, 1981–2003).

——, *De non reddenda Kallipoli petente Amurate, PG* 154, cols. 1009–1036.

——, *Oratio pro subsidio Latinorum, PG* 156, cols. 962–1008.

Lampros, S. *Παλαιολόγεια και Πελοποννησιακά* [*Palaiologeia kai Peloponnesiaka*] 4 vols. (Athens, 1912–1930).

Laskaris, Theodore II, *Theodori Ducae Laskaris Epistulae CCXVII*, N. Festa (ed.), (Florence, 1898).

——, *Κοσμική Δήλωσις* [*Kosmike Delosis*], N. Festa (ed.), *Giornale della società Asiatica Italiana*, 11 (1897–1898), 97–113 (first book).

——, *Encomio dell' imperatore Giovanni Duca*, L. Tartaglia (ed.), (Napoli, 1990).

——, *Opuscula rhetorica*, L. Tartaglia (ed.), (Munich, 2000).

Leo the Deacon, *Leonis diaconis Caloensis Historiae libri decem*, C.B. Hase (ed.), (Bonn, 1828); tr. A.-M.Talbot - D.F. Sullivan, *The History of Leo the Deacon* (Washington, 2005).

Les chroniques de la ville de Metz, J. Huguenin (ed.), (Metz, 1835).

Libro de los fechos et conquistas del principado de la Morea, A.M. Fatio (ed.), (Geneve, 1885).

Livre de la conqueste de la Princée de l Amoré, J. Longnon (ed.), (Paris, 1911).

Magdalino, P. "An Unpublished *pronoia* Grant from the Second Half of the Fourteenth Century," *ZRVI* 18 (1978), 155–163.
Magistros, Thomas, *Oratio de regis oficiis ad Andronicum II, PG* 145, cols. 495–548.
——, *Oratio de subditorum officiis, PG* 145 cols. 449–495.
Matthew of Ephesos, *Die Briefe des Matthaios von Ephesos im Codex Vindobonensis Theol. Gr. 174*, D. Reinsch (ed.), (Berlin, 1974).
Maurice, *Das Strategikon des Maurikios*, G.T. Dennis (ed.), (Vienna, 1981).
Mazaris' Journey to Hades: or Interviews with Dead Men About Certain Officials In the Imperial Court, J.N. Barry *et als.* (ed.), (Buffalo, 1975).
Melville Jones, J.-R. *The Siege of Constantinople 1453. Seven Contemporary Accounts* (Amsterdam, 1973).
Melikoff-Sayar, I. (ed.), *Le Destan d' Umur Pascha* (Paris, 1957).
Menander Rhetor, D. Rusell - N. Wilson (ed.), (Oxford 1981).
McGeer, E. *Sowing the Dragon's Teeth: Byzantine Warfare in the Tenth Century* (Washington, 1995).
Mercati, G. *Notizie a Procoro e Demetrio Cidone, Manuele Caleca e Teodoro Meliteniota* (Vatican, 1930).
——, *Collectanea Byzantina* (Bari, 1970).
Metochites, Theodore, *Miscellanea philosophica et historica*, G. Müller (ed.), (Leipzig, 1821).
——, *Οι δύο Βασιλικοί Λόγοι* [*The Two Imperial Orations*], I. Polemis (ed.), (Athens, 2007).
Miklosich, F. - Müller, J. (eds.), *Acta e diplomata Graeca medii aevi sacra et profana*, 6 vols. (Vienna, 1860–1890).
Muntaner, Ramon, *Crònica*, 2 vols., V.J. Escartí (ed), (Valencia, 1999).
Oikonomides, N. "Cinq actes inèdites du patriarche Michel Autoreianos," *REB* 25 (1967), 113–145.
Palaiologos, Manuel II, *Συμβουλευτικός πρός τούς Θεσσαλονικεῖς*, V. Laourdas (ed.), *Makedonika* 3 (1953–1955), 295–303.
——, *The Letters of Manuel II Palaeologus*, G.T. Dennis (ed.), (Washington, 1977).
——, *Funeral Oration on his brother Theodore*, J. Chrysostomides (ed.), (Thessalonica, 1985).
Palaiologos, Michael VIII, *Imperatoris Michaelis Palaeologi de vita sua*, H. Grégoire (ed.), *B* 29–30 (1959–1960), 448–474.
Palaiologos, Theodore, marquis of Montferrat, *Les enseignements de Théodore Paléologue*, C. Knowles (ed.), (London, 1983).
Perria, L. "Due documenti greci del XIV secolo in un codice della biblioteca vaticana (vat. gr. 1335 con un tavole)," *JÖB* 30 (1981), 259–297.
Pertusi, A. *La Caduta di Costantinopoli*, 2 vols. (Rome, 1976).
Philes, Manuel, *Manuelis Philae carmina*, 2 vols. E. Miller (ed.), (Paris, 1855–1857).
Planoudes, Maximos, "Le basilikos de Maximos Planoudes," L.G. Westernik (ed.), *BSI* 27 (1966) 98–103, 28 (1967), 54–67, 29 (1968), 34–50.
——, *Maximi Monachi Planudis Epistulae*, A. Leone (ed.), (Amsterdam, 1991).
Prinzig, G. "Der Brief Kaiser Heinrichs von Konstantinopel vom 13. Januar 1212," *B* 43 (1973), 395–431.
Pseudo-Kodinos, *Traité des offices*, J. Verpeaux (ed.), (Paris, 1966).
Sanudo, Marino Torsello, *Marin Sanudo Torsello. Istoria di Romania*, E. Papadopoulou (ed.), (Athens, 2000).
Schreiner, P. "Zwei unedierte Praktika aus der zeiten Hälfte des 14. Jahrhunderts," *JÖB* 19 (1970), 33–49.
——, *Die Byzantinischen Kleinchroniken*, 3 vols. (Vienna, 1975–1979).
Sylloge tacticorum quae olim 'inedita Leonis Tactica' dicebatur, A. Dain (ed.), (Paris, 1938).
Sphrantzes, *Cronaca*, R. Maisano (ed.), (Rome, 1990).

Synopsis Chronike, K. Sathas (ed.), *Μεσαιωνική Βιβλιοθήκη VII*, 7 vols. (Athens, 1898).
Thalloczy L. - Jerejek, C. (eds.), *Acta et diplomata Res Albaniae Mediae Aetatis Illustrantia* (Vienna, 1913).
Theognostos, *Theognosti Thesaurus*, J. Munitiz (ed.), (Brepols, 1979).
Thiriet, F. *Régestes des délibérations du Sénat de Venise concernant la Romanie*, 3 vols. (Paris, 1958–1961).
——, *Délibérations des assemblées vénitiennes concernant la Romanie*, 2 vols. (Paris, 1966–1971).
The Byzantine Achilleid. The Naples version, O.L. Smith (ed.), (Vienna, 1999).
The War of Troy, Papathomoupoulou M. - Jeffreys E. (eds.), (Athens, 1996).
Thomas, J. - Hero, A.C. (ed.), *Byzantine Monastic Foundation Documents*, 5 vols. (Washington, 2000).
Villehardouin, Geoffroy de, *La Conqueste de Constantinople*, A. Pauphilet (ed.), *Historiens et chroniqueurs du Moyen Age* (Paris, 1952).
y Lluch, A.R. *Diplomatari de l'Orient Català (1301–1409)* (Barcelona, 1947 repr. 2001).
Zepos, I. *Jus Graecoromanum*, 8 vols. (Athens 1931, reprint Aalen 1962).

Secondary Sources

Aerts, W.J. - Hokwerda, H. *Lexikon of the Chronicle of the Morea* (Groningen, 2002).
Ágoston, G. *Guns for the Sultan. Military Power and the Weapons Industry in the Ottoman Empire* (Cambridge, 2005).
Ahrweiler, H. "La politique agraire des empereurs de Nicée," *B* 29 (1958), 51–66.
——, "L'histoire et la géographie de la région de Smyrne entre les deux occupations Turques (1081–1317) particulièrement au XIII siècle," *TM* 1(1965), 1–204.
——, *Byzance et la mer* (Paris, 1966).
Angelov, D. "Byzantine Imperial Panegyric as Advice Literature," in E. Jeffreys (ed.), *Rhetoric in Byzantium* (Aldershot, 2003), 59–73.
——, "Byzantine ideological reactions to the Latin conquest of Constantinople," in A. Laiou (ed.), *Urbs Capta. The Fourth Crusade and its Consequences* (Paris, 2005), 293–310.
——, *Imperial Ideology and Political Thought in Byzantium. 1204–1328* (Cambridge, 2007).
Angold, M. *A Byzantine Government in Exile. Government and Society under the Laskarids of Nicaea 1204–1261* (Oxford, 1974).
Argenti, Ph. *The Occupation of Chios by the Genoese*, 3 vols. (Cambridge, 1958).
Asdracha, C. *La région des Rhodopes aux XIIIe et XIVe siècles: étude de géographie historique* (Athens, 1976).
Asutay-Effenberger, N. *Die Landmauer von Konstantinopel-Istanbul. Historisch-Topographische und baugeschichtliche Untersuchungen* (Berlin, 2007).
Bachrach, B.S. "Caballus et caballarius in Medieval Warfare," in H. Chikering - T.H. Seiler (eds.), *The Study of Chivalry. Resources and Approaches* (Kalamazoo, 1988), 173–211.
Balard, M. "À propos de la bataille du Bosphore. L'expedition genoisie de Paganino Doria à Constantinople 1351–1352," *TM* 4 (1970), 431–470.
——, *La Romanie génoise: XIIe-début du XVe siècle*, 2 vols. (Rome, 1978).
Barber, R. *The Reign of Chivalry* (London, 1980).
Barber, R. - Barker, J. *Tournaments, Jousts Chivalry and Pageants in the Middle Ages* (Woodbridge, 1989).
Barker, J.W. *Manuel II Palaeologus (1391–1425): A study in Late Byzantine Statemanship* (New Brunswick, 1969).

——, "The Problem of Appanages in Byzantium during the Palaiologan Period," *Byzantina* 3 (1971), 105–123.
Bartusis, M. "State Demands for the Billeting of Soldiers in Late Byzantium," *ZRVI* 26 (1987), 115–123.
——, "The Kavallarioi of Byzantium," *Speculum* 63 (1988), 343–351.
——, "Urban Guard Service in Late Byzantium: The Terminology and the Institution," *Macedonian Studies* 2 (1988), 52–77.
——, "The Megala Allagia and the Tzaousios: Aspects of Provincial Military Organization in Late Byzantium," *REB* 47 (1989), 183–205.
——, "On the Problem of Smallholding Soldiers in Late Byzantium," *DOP* 44 (1990), 1–26.
——, *The Late Byzantine Army. 1204–1453* (Philadelphia, 1992).
Beaton, R. *The Medieval Greek Romance* (Cambridge, 1986).
Beldiceanu, N. *Le timar dans l'État ottoman: début XIVe–début XVIe siècle* (Wiesbaden, 1987).
Beldiceanu-Steinherr, I. "La conquête d' Adrianople par les Turcs: La penetration Turque en Thrace et la valeur des chroniques Ottomans," *TM* 1 (1965) 439–461.
——, "En marge d'un acte concernant le penğyek et les aqinği," *REI* 38 (1969), 21–47.
——, "L' installation des Ottomans," B. Geyer – J. Lefort (eds.), *La Bithynie au moyen âge* (Paris, 2003), 351–374.
Belting, H. – Mango, C. – Mouriki, D. *The Mosaics and Frescoes of St Mary of Pammakaristos (Fethiye Camii) at Istanbul* (Washington, 1978).
Birkenmeier, J. *The Development of the Komnenian Army. 1081–1180* (Leiden, 2002).
Bliese, R.E. "Rhetoric and Morale: A Study of Battle Orations from the Central Middle Ages," *JMH* 15 (1989), 201–226.
Blum, W. *Georgios Gemistos Plethon. Politik, Philosophie und Rhetorik im Spätbyzantinischen Reich* (Stuttgart, 1988).
Bombaci, A. "The Army of the Seljuqs of Rum," *Istituto Orientale di Napoli, Annali*, 38 (1978), 343–369.
Bon, A. *La Morée franque (1205–1430)* (Paris, 1969).
Bosch, U. *Kaiser Andronikos III Palaiologos. Versuch einer Darstellung der Byzantinischen Geschichte in den Jahren 1321–1341* (Amsterdam, 1965).
Browing, R. "Greeks and Others: From Antiquity to the Renaissance," in T. Harrison (ed.), *Greeks and Barbarians* (Edinburgh, 2002), 257–277.
Bryer, A. "Greek Historians on the Turks: the Case of the First Byzantine-Ottoman Marriage," in R.H.C. Davis – J.M.H. Wallace-Hadrill (eds.), *The Writing of History in the Middle Ages. Essays presented to Richard William Southern*, (Oxford, 1981), 471–493.
Cahen, C. *Pre-Ottoman Turkey*, tr. J. Jones-Williams (London, 1968).
Calder, N. "Khums in Imami Shi'I Jurisprudence, from the Tenth to the Sixteenth Century A.D.," *Bulletin of the School of Oriental and African Studies* 45 (1982), 39–47.
Carlson, D. "Church councils on Chivalry," in H. Chikering – T.H. Seiler (eds.), *The Study of Chivalry. Resources and Approaches* (Kalamazoo, 1988), 141–171.
Caffero, W. *Mercenary Companies and the Decline of Siena* (Baltimore, 1998).
——, " Slaying the Hydra-Headed Beast: Italy and the Companies of Adventure on the Fourteenth Century," in L.J.A. Villalon – D.J. Kagay (eds.), *Crusaders, Condottieri and the Cannon. Medieval Warfare in Societies around the Mediterranean* (Leiden, 2003), 286–304.
——, *John Hawkwood. An English Mercenary in Fourteenth-Century Italy* (Baltimore, 2006).
Chevedden, P.E. "The Invention of the Counterweight Trebuchet: A Study in Cultural Diffusion," *DOP* 54 (2000), 71–116.

Contamine, Ph. "Les tournois en France à la fin du moyen âge," in J. Fleckenstein (ed.), *Das ritterliche Turnier im Mittelalter* (Göttingen, 1985).
Connely, S. "Fortifying the City of God: Dardanus' Inscription Revisited," *Classical Journal* 102 (2007), 145–154.
Cox, E.L. *The Green Count of Savoy, Amadeus VI and Transalpine Savoy in the Fourteenth Century* (Princeton, 1967).
Crouch, D. "William Marshall and the Mercenariat," in J. France (ed.), *Mercenaries and Paid Men. The Mercenary Identity in the Middle Ages* (Leiden, 2008), 16–32.
Ćurčić, S. "Architecture in the Age of Insecurity. An Introduction to Secular Architecture in the Balkans," in S. Ćurčić – E. Hadjitryphonos (eds.), *Secular Medieval Architecture in the Balkans, and its Preservation. 1300–1500* (Thessalonica, 1997), 19–51.
——, "Late Medieval Fortified Palaces in the Balkans. Security and Survival," *Mnemeio kai Perivallon*, 6 (2000), 11–48.
——, *Architecture in the Balkans from Diocletian to Süleyman* (Yale, 2010).
Dain, A. "La partage du butin de guerre d' après les traités juridiques et militaires," in *Actes du VI*[e] *congrès d' etudes byzantines*, I (Paris, 1950), 347–352.
Dain, A. – de Foucault, J.-A. "Les stratégistes Byzantins," *TM* 2 (1967), 317–392.
Delbrück, H. *History of the Art of War, volume III. Medieval warfare*, tr. W.J. Renfroe (London, 1990).
Dennis, G.T. *The Reign of Manuel II Palaeologus in Thessalonica, 1382–1387* (Rome, 1960).
——, "The Capture of Thebes by the Navarrese (6 March 1378) and other Chronological Notes," *OCP* 26 (1960), 41–50.
——, "The Byzantines in Battle," in N. Oikonomides (ed.), *Byzantium at War (9th–12th c.)* (Athens, 1997), 165–178.
——, "Imperial Panegyric: Rhetoric and Reality," in H. Maguire (ed.), *Byzantine Court Culture from 829 to 1204* (Washington, 1997), 131–140.
——, "Byzantine Heavy Artillery: The Helepolis," *GRBS* 39 (1998), 99–115.
De Vries, K. "Gunpowder Weapons at the siege of Constantinople in 1453," in L. Yaakov (ed.), *War Armies and Soldiery in the Eastern Mediterranean* (Leiden, 1986), 343–362.
——, *Medieval Military Technology* (New York, 1992).
——, "The Impact of Gunpowder Weaponry on Siege Warfare in the Hundred Years War," in I.A. Corfis – M.Wolff (ed.), *The Medieval City under Siege* (Woodbridge, 1995).
——, *Infantry Warfare in the Early Fourteenth Century* (New York, 1996).
Di Cosmo, N. "Inner Asian ways of Warfare in Historical Perspective," in N. di Cosmo (ed.), *Warfare in Inner Asian History* (Leiden, 2002), 1–29.
Dragon, G. "Byzance et le modèle islamique au Xe siècle, à propos des Constitutions Tactiques de l' empereur Léon VI," in *Comptes rendus des séances de l' Académie des Inscriptions et Belles-Lettres* (Paris, 1983), 219–243.
Dunbabin, J. *Charles I of Anjou* (London/New York, 1998).
Engels, D.N. *Alexander the Great and the Logistics of the Macedonian Army* (Los Angeles, 1978).
Failler, A. "La restauration et la chute definitive de Tralleis au 13e siècle," *REB* 42 (1984), 249–263.
——, "Chronologie et composition dans l' Histoire de George Pachymérès (livres VII–XII)," *REB* 48 (1990), 5–87.
——, "Pachymeriana alia," *REB* 51 (1993), 237–260.
——, "Les Émirs Turcs à la conquête de l'Anatolie au début du 14e siècle," *REB* 52 (1994), 69–112.
Fine, J.V.A. *The Late Byzantine Balkans* (Kalamazoo, 1995).
Fögen, M.T. "Die Scholien zur Hexabiblos im Codex vetustissimus Vaticanus Ottobonianus," *FM* 4 (1981), 268–275.

Foss, C. *Byzantine and Turkish Sardis* (Harvard, 1979).
——, "Byzantine Malagina and the Lower Sangarius," *Anatolian Studies* 40 (1990), 161–183.
——, *Survey of Medieval Castles of Anatolia II. Nikomedia* (Ankara, 1996).
Foss, C. - Winfield, D. *Byzantine Fortifications. An Introduction* (Praetoria, 1986).
France, J. *Victory in the East. A Military History of the First Crusade* (Cambridge, 1994).
——, *Western Warfare in the Age of the Crusades. 1000–1300* (London, 1999).
——, "Mercenaries and Paid Men. The Mercenary Identity in the Middle Ages. Introduction," in J. France (ed.), *Mercenaries and Paid Men. The Mercenary Identity in the Middle Ages*, (Leiden, 2008), 1–13.
Frankopan, P. "Land and Power in the Middle and Later Period," in J. Haldon (ed.), *A Social History of Byzantium* (Oxford, 2009), 112–142.
Geanakoplos, D. *Emperor Michael Palaiologos and the West. A Study in Byzantine-Latin Relations.1258–1282* (Harvard, 1959).
Giannakopoulos, D.-K. *Δουκάτο των Αθηνών. Η κυριαρχία των Acciaiuioli* [*The Duchy of Athens. The Domination of the Acciaiuioli*] (Thessalonica, 2006).
Giros, C. "Les fortifications médiévales," in B. Geyer - J. Lefort (eds.), *La Bithynie au Moyen âge* (Paris, 2003), 210–224.
Gounaridis, P. *Το κίνημα των Αρσενιατών (1261–1310)* [*The Movement of the Arsenites*] (Athens, 1999).
Goutziokostas, A. "*Ο κριτής του στρατοπέδου και ο κριτής του φοσσάτου*," [The *krites tou stratopedou* and the *krites tou phossatou*] *Byzantina* 26 (2006), 79–99.
Guilland, R. *Recherches sur les institutions Byzantines*, 2 vols. (Amsterdam, 1967).
Haldon, J. *Byzantine praetorians* (Bonn, 1984).
——, "Limnos, Monastic Holdings and the Byzantine State," in A. Bryer - H. Lowry (eds.), *Continuity and Change in Late Byzantine and Early Ottoman Society* (Birmingham, 1986), 161–188.
——, *The State and the Tributary Mode of Production* (London, 1993).
——, "Blood and Ink: Some Observations on Byzantine Attitudes Towards Warfare and Diplomacy," in J. Shepard - S. Franklin (eds.), *Byzantine Diplomacy* (London, 1994), 281–294.
——, *Warfare State and Society in the Byzantine World. 565–1204* (London, 1999).
——, "*The krites tou stratopedou*: a new Office for a new Situation?' in C. Morrison (ed.), *Mélanges Gilbert Dagron*," *TM* 14 (2002), 279–286.
——, "Approaches to an Alternative Military History of the Period ca. 1025–1071," in A. Avramea - A. Laiou - E. Chryssos (eds.), *The empire in Crisis (?) Byzantium in the 11th Century (1025–1081)* (Athens, 2003), 45–74.
——, *The Palgrave Atlas of Byzantine History* (New York, 2005).
——, "Social Élites, Wealth and Power," in J. Haldon (ed.), *A Social History of Byzantium* (Oxford, 2009), 168–211.
Haldon, J. - Kennedy, H. "The Arab-Byzantine Frontier in the Eighth and Ninth Centuries: Military Organisation and Society in the Borderlands," *ZRVI* 19 (1980), 79–116.
Halecki, O. *Un Empereur de Byzance à Rome* (Warsaw, 1930 repr. London, 1972).
Hannick, C. - Schmalzbauer, G. "Die Synadenoi: Prosopographische Untersuchung zu einer byzantinischen Familie," *JÖB* 25 (1976), 125–161.
Held, J. *Hunyadi. Legend and Reality* (New York, 1995).
Hendrickx, B. *Οι πολιτικοί και στρατιωτικοί θεσμοί της Λατινικής Αυτοκρατορίας της Κωνσταντινουπόλεως κατά τους πρώτους χρόνους της υπάρξεως της* [*The Political and Military Institutions of the Latin Empire of Constantinople in the First Years of its Existence*] (Thessalonica, 1970).
——, "Allagion, Tzaousios et Protallagatôr dans le context Moréote: quelques remarques," *REB* 50 (1992), 208–217.

Hendy, M. *Studies in the Byzantine Monetary Economy c. 300–1450* (Cambridge, 1985).

Hero, A.C. "Theoleptos of Philadelpheia (ca. 1250–1322): From solitary to Activist," in S. Ćurčić – D. Mouriki (eds.), *The Twilight of Byzantium* (Princeton, 1991), 27–38.

Heywood, K. "Low-level diplomacy between Byzantines and Ottoman Turks: the Case of Bythinia," in J. Shepard – S. Franklin (eds.), *Byzantine Diplomacy* (London, 1994), 51–55.

——, "Notes on the Production of Fifteenth-Century Ottoman Cannon," *Proceedings of the International Symposium on Islam and Science*, (Islamabad, 1981) reprinted in *Writing Ottoman History* (Aldershot, 2002), 1–5.

Hillenbrand, C. *The Crusades. Islamic Perspectives* (Edinburgh, 1999).

Hohlweg, A. *Beiträge zur Verwaltungsgeschichte des Oströmischen Reiches unter den Komnenen* (Munich, 1965).

Housley, N. "One Man and His Wars: The Depiction of Warfare by Marshall Boucicaut's Biographer," *JMH* 29 (2003), 27–40.

Hunger, H. "Un roman byzantin: Callimaque et Chrysorroe et son atmosphère," *TM* 3 (1968), 405–442.

——, *Die hochsprachliche profane Literatur der Byzantiner*, 2 vols. (Munich, 1978).

——, "Urkunden–und Memoirtext: Der Chrysobullos Logos des Johannes Kantakuzenos für Johannes Angelos," *JÖB* 27 (1978), 107–125.

Hyland, A. *The Medieval Warhorse from Byzantium to the Crusades* (London, 1994).

Ilieva, A. *Frankish Morea 1205–1262. Socio-Cultural Interaction Between the Franks and The Local Population* (Athens, 1991).

Imber, C. *The Ottoman Empire 1300–1600. The Structure of Power* (London, 2002).

İnalcık, H. "Ottoman Methods of Conquest," *Studia Islamica* 2 (1954), 104–129.

——, "Stefan Duşan' dan Osmanlı imperatorluğuna," [From Stefan Dušan to the Ottoman Empire] in H. İnalcık (ed.), *Fatih devri üzerinde tetkiler ve vesikalar* [Researches and Documents on the Period of the Conqueror] (Ankara, 1954), 137–225.

——, *Gelibolu*, *EI*², cols. 983–987.

——, "The conquest of Edirne (1361)," *Archivum Ottomanicum* 3 (1971), 185–210.

——, "Osman Ghazi's siege of Nicaea and the battle of Bapheus", in E. Zachariadou (ed.), *The Ottoman Emirate. 1300–1389* (Rethymnon, 1993), 77–99.

——, "How to Read Ashik Pasha-Zade's History," in C. Imber – C. Heywood (eds.), *Studies in Ottoman History in Honour of Professor V.L. Mènage* (Istanbul, 1994), 141–144.

Ivison, E. "Urban Renewal and Imperial Revival in Byzantium (730–1025)," *BF* 26 (2000), 1–43.

Jackson, P. *The Mongols and the West 1221–1410* (London, 2005).

Jacoby, D. "Les Vénitiens naturalisés dans l' Empire byzantine: un aspect de l' expansion de Venise en Romanie du XIIIe au milieu du XVe siècle," *TM* 8 (1981), 217–235.

——, "Knightly Values and Class Consciousness in the Crusader States of the Eastern Mediterranean," *MHR* 1 (1986), 158–186.

——, "Foreigners and the Urban Economy in Thessalonike c. 1150–c.1450," *DOP* 57 (2003), 85–132.

Jimenez, M.G. "Frontier and Settlement in the Kingdom of Castile (1085–1350)," in R. Bartlett – A. Mackay (eds.), *Medieval Frontier Societies* (Oxford, 1989), 49–74.

Jocović, N. – Nešković, J. "Fortifications of Smederevo," in S. Ćurčić – E. Hadjitryphonos (eds.), *Secular Medieval Architecture in the Balkans, 1300–1500, and its Preservation* (Thessalonica, 1997), 132–135.

Kafadar, C. *Between Two Worlds. The Construction of the Ottoman State* (Los Angeles 1995).

Kaegi, W. "The Contribution of Archery to the Conquest of Anatolia," *Speculum* 39 (1964), 96–108.

Karlin-Hayter, P. "Notes sur le ΛΑΤΙΝΙΚΟΝ dans l' armée et les historiens de Nicée," *BF* 4 (1972), 142–150.

Katsoni, P. "*Ο χρόνος έναρξης της Βυζαντινής υποτέλειας στους Οθωμανούς*," [The Time of the Beginning of the Byzantine Vassalage to the Ottomans] *Byzantiaka* 14 (1994), 459–481.

——, *Μια επταετία κρίσιμών γεγονότων. Το Βυζάντιο στα έτη 1366–1373* [*Seven Years of Crucial Events. The Byzantine Empire in the Period 1366–1373*] (Thessalonica, 2002).

Kaeuper, R. *Chivalry and Violence in Medieval Europe* (Oxford, 1999).

Kazhdan, A. "The Aristocracy and the Imperial Ideal," in M. Angold (ed.), *The Byzantine Aristocracy IX to XII Centuries* (Oxford, 1984), 43–57.

——, *The Oxford Dictionary of Byzantium* (Oxford, 1991).

——, "State, Feudal and Private Economy in Byzantium," *DOP* 47 (1993), 83–100.

——, "*Pronoia*: The History of a Scholarly Discussion," *MHR* 10 (1995), 133–163.

——, "The Italian and Late Byzantine City," *DOP* 49 (1995), 1–22.

Kazhdan, A. - Epstein, A.W. *Change in Byzantine Culture in the Eleventh and Twelfth Centuries* (Berkeley/London, 1985).

Keen, M. *The Laws of War in the Late Middle Ages* (London, 1965).

——, *Chivalry* (Yale, 1984).

Khaddouri, M. *War and Peace in the Laws of Islam* (Baltimore 1955).

Kianka, F. *Demetrius Kydones (1324–1397). Intellectual and Diplomatic Relations Between Byzantium and the West in the Fourteenth Century* (Reston, 1981).

Klopf, M. "The Army in Constantinople at the Accession of Constantine XI," *B* 40 (1970), 385–392.

Kiousopoulou, A. *Βασιλεύς ή οικονόμος; Πολιτική εξουσία και ιδεολογία πριν την άλωση* [*Emperor or Administrator? Political Power and Ideology Before the Fall*] (Athens, 2007).

Kolia-Dermitzaki, A. *Ο Βυζαντινός Ιερός Πόλεμος* [*The Byzantine Holy War*] (Athens, 1991).

Kolias, G. "Η ανταρσία του Ιωάννη Ζ' Παλαιολόγου εναντίον του Ιωάννη Ε' Παλαιολόγου (1390)," [The Revolt of John VII Palaiologos Against John V Palaiologos] *Hellenika* 12 (1952), 34–64.

Kollias, T. "Essgewohnheiten und Verpflegung im Byzantinischen Heer," in W. Hörandner *et al.* (eds.), *Byzantios.* Festschrift für Herbert Hunger zum 70. Geburtstag," (Vienna, 1984), 193–202.

——, "The *Taktika* of Leo VI the Wise and the Arabs," *Graeco-Arabica* 3 (1984), 129–135.

——, "Πολεμική τακτική των Βυζαντινών. Θεωρία και πράξη," [Battlefield Tactics of the Byzantines. Theory and Practice] in N. Oikonomides (ed.), *Byzantium at War (9th–12th c.)* (Athens, 1997), 154–164.

——, "Ein zu wenig bekannter Faktor im byzantinischen Heer. Die Hilfkräfte," in C. Scholz (ed.), *ΠΟΛΥΠΛΕΥΡΟΣ ΝΟΥΣ. Miscellanea für Peter Schreiner zu seinem 60. Geburtstag* (Munich/Leipzig, 2000), 113–124.

Kyriakidis, S. "The Role of the *megas domestikos* in the Late Byzantine Army (1204–1453)," *BSl* 66 (2008), 241–258.

——, "The Division of Booty in Late Byzantium. 1204–1453," *JÖB* 58 (2009), 164–176.

——, "The Employment of Large Groups of Mercenaries in Byzantium in the Period ca. 1290–1305 as Viewed by the Sources," *B* 69 (2009), 208–230.

——, "Byzantine Responses to the Battlefield Tactics of the Turkoman Principalities: The Battle of Pelekanos," *BZ* 103 (2010), 83–97.

Kyritses, D. *The Byzantine Aristocracy in the Thirteenth and Fourteenth Centuries*, unpublished PhD Diss. (Harvard, 1997).

——, "The 'Common Chrysobulls' of Cities," *VyzSym* 13 (1999), 229–245.

——, "Κράτος και Αριστοκρατία στην εποχή του Ανδρνίκου Β'. Το αδιέξοδο της στασιμότητας," [State and Aristocracy in the Time of Andronikos II. The Stalmate of Stagnation] in L. Maurommatis (ed.), *Ο Μανουήλ Πανσέληνος και η εποχή του* [*Manuel Panselenos and his Time*] (Athens, 1999), 177–194.

——, "Η Άλωση της Κωνσταντινούπολης και το τέλος του Βυζαντινού πολιτισμού," [The Fall of Constantinople and the End of Byzantine Civilization] in T. Kiousopoulou (ed.), *Η Άλωση της Κωνσταντινούπολης και η μετάβαση από τους μεσαιωνικούς στους νεότερους χρόνους* [*The Fall of Constantinople and the Transition from the Medieval to Modern Times*] (Rethymnon, 2005), 161–172.

Laiou, A. *Constantinople and the Latins. The Foreign Policy of Andronikos II* (Cambridge, MA 1972).

——, "The Byzantine Aristocracy in the Palaeologan Period: a Story of Arrested Development," *Viator* 4 (1973) 131–151.

——, *Peasant Society in the Late Byzantine Empire: A Social and Demographic Study* (Princeton, 1977).

——, "Some Observations on Alexios Philanthropenos and Maximos Planoudes," *BMGS* 4 (1978), 89–99.

——, "On Just War in Byzantium," in S. Reinher - J. Langdon (ed.), 2 vols. *TO ΕΛΛΗΝΙΚΟΝ. Studies in Honor of Spyros Vryonis*, (New York, 1993), 155–177.

——, "Italy and the Italians in the Political Geography of the Byzantines (14th Century)," *DOP* 49 (1995), 73–98.

Lampakes, S. *Γεώργιος Παχυμέρης. Πρωτέκδικος και δικαιοφύλαξ* [*George Pachymeres. Protekdikos and dikaiophylax*] (Athens, 2004).

Lampe, G.W.H. *A Patristic Greek Lexikon* (Oxford, 1961).

Langdon, J. *John III Ducas Vatatzes' Byzantine Imperium in Anatolian Exile, 1222–54* (Los Angeles, 1978).

——, *Byzantium's last Imperial Offensive in Asia Minor* (New Rochelle, 1992).

Lemerle, P. *L Emirat d'Aydin, Byzance et l'Occident. Recherches sur la geste d'Umur Pascha* (Paris, 1957).

Lock, P. *The Franks in the Aegean. 1204–1500* (London/New York, 1995).

Loenertz, R. "Hospitaliers et Navarrais en Grèce 1376–1383. Régestes et documents," *OCP* 22 (1956), 319–360.

Longnon, J. *L' empire Latin de Constantinople et la principauté de Morée* (Paris, 1949).

Lowry, H. *The Nature of the Early Ottoman State* (New York, 2003).

——, *The Shaping of the Ottoman Balkans. The Conquest, Settlement and Infrastructural Development of Northern Greece* (Istanbul, 2008).

Luttrell, A. "Appunti sulle compagnie navarresi in Grecia," *Rivista di studi bizantini e slavi* 3 (1983), 113–127.

Macrides, R. "The New Constantine and the new Constantinople-1261?" *BMGS* 6 (1980) 13–41.

——, "From the Komnenoi to the Palaiologoi: Imperial Models in Decline and Exile," in P. Magdalino (ed.), *New Constantines: The Rhythm of Imperial Renewal in Byzantium, 4th–13th Centuries* (Aldershot, 1994), 269–282.

——, "The Thirteenth Century in Byzantine Historical Writing," in C. Dedrinos *et als.* (eds.), *PORPHYROGENITA. Essays on the History and Literature of Byzantium and the Latin East in Honour of Julian Chrysostomides* (London, 2002), 63–76.

——, "The Reason is not Known. Remembering and Recording the Past. Pseudo-Kodinos as a Historian," in P. Odorico - P. Agapitos - M. Hinterberger (eds.), *L' écriture de la mémoire. La littérarité de la historiographie* (Paris, 2006), 317–330.

——, *George Akropolites. The History* (Oxford, 2007).

Magdalino, P. "Between Romaniae: Thessaly and Epiros in the Late Middle Ages," in B. Arbel - D. Jacoby (eds.), *Latins and Greeks in the Eastern Mediterranean after 1204* (London, 1989), 87–104.

——, *The Empire of Manuel I Komnenos. 1118–1143* (Cambridge, 1993).

Maguire, H. - Jones, L. "A Description of the Jousts of Manuel I Komnenos," *BMGS* 26 (2002), 104–148.

Majeska J. *Russian Travelers to Constantinople in the Fourteenth and Fifteenth Centuries* (Washington, 1984).
Mango, C. - Hawkins, E.J.W. "Additional Finds at Fenari Isaa Camii," *DOP* 22 (1968), 177–184.
Maksimovič, L. *The Byzantine Provincial Administration under the Palaiologoi* (Amsterdam, 1988).
——, "Η ανάπτυξη κεντρόφυγων ροπών στις σχέσεις Βυζαντίου-Σερβίας τον 14° αιώνα," [The Development of Centrifugal Tendencies in Byzantine-Serbian Relations in the Fourteenth Century] in N. Oikonomides (ed.), *Byzantium at War* (Athens, 1997), 283–290.
Mallett, M. *Mercenaries and their Masters* (London, 1974).
——, "Condottieri and Captains in Renaissance Italy," in D.J.B. Trim (ed.), *The Chivalric Ethos and the Development of Military Professionalism* (Leiden, 2003), 67–88.
Mallett, M. - Hale, J.R. *The Military Organization of a Renaissance State. Venice c.1400 to 1617* (Cambridge, 1984).
Marshall, C. *Warfare in the Latin East 1102–1291* (Cambridge, 1992).
Matschke, K.-P. *Fortschritt und Reaktion in Byzanz im 14. Jahrhundert: Konstantinopel in der Bürgerkriegsperiode von 1341 bis 1354* (Berlin, 1971).
——, "Johannes Kantakuzenos, Alexios Apokaukos und die byzantinische Flotte in der Burgerkriegsperiode, 1340–1355," in *Actes du XIVe Congrès international des études byzantines* (Bucarest, 1971), 193–205.
Matschke, K.-P. - Tinnefeld, F. *Die Gesellschaft im späten Byzanz* (Koln/Weimar/Vienna, 2001).
Mavromatis, L. *La fondation de l' Empire Serbe. Le Kralj Milutin* (Thessalonica, 1978).
McGeer, E. "Byzantine Siege Warfare in Theory and Practice," in I.A. Corfis - M. Wolfe (eds.), *The Medieval City under Siege* (Woodbridge, 1995), 123–129.
McNeal-Caplow, H. "Michelozzo at Ragusa: New Documents and Revaluations," *Journal of The Society of Architectural Historians* 31/2 (1972), 108–119.
Mélikoff, I. *Germiyān-Oghulları*, in *EI*², cols. 988–990.
Milhajlovski, R. "The Battle of Pelagonia, 1259: a New Look at the March Routes and Topography," *BSI* 64 (2006), 275–284.
Moravscik, G. *Bizantinoturcica*, 2 vol. (Berlin, 1958).
Morillo, P. "The 'Age of Cavalry' revisited," in D.J. Karay - L.J.A. Villalon (eds.), *The Circle of War in the Middle Ages. Essays on Medieval Military and Naval History* (Woodbridge, 1999).
Moutsopoulos, N.-K. *Ρεντίνα*, [*Rentina*] 4 vols. (Thessalonica/Athens, 1995–2001).
——, *Άγνωστα Βυζαντινά κάστρα της Μακεδονίας* [*Unknown Byzantine Castles of Macedonia*] (Thessalonica, 2004).
Munitiz, J. "War and Peace Reflected in Some Byzantine Mirrors of *Princes*," in T.S. Miller - J. Nesbitt (eds.), *Peace and War in Byzantium: Essays in Honor of George T. Dennis* (Washington, 1995), 51–61.
Murphy, R. *Ottoman Warfare. 1500–1700* (London, 1999).
Necipoğlu, N. "Social and Economic Conditions in Constantinople during Mehmed II's Siege," in T. Kiousopoulou (ed.), *Η Άλωση της Κωνσταντινούπολης και η μετάβαση από τους μεσαιωνικούς στους νεώτερους χρόνους* [The Capture of Constantinople and the Transition from the Medieval to the Modern Times] (Herakleion, 2005), 75–86.
——, *Byzantium Between the Ottomans and the Latins* (Cambridge, 2009).
Nicol, D. *The despotate of Epiros* (Oxford, 1957).
——, *The Byzantine Family of Kantakouzenos (Cantacuzenus) ca. 1100–1460* (Washington, 1968).
——, *The Last Centuries of Byzantium* (London, 1972).
——, *The despotate of Epiros. 1267–1479. A Contribution to the History of Greece* (Cambridge, 1984).

——, *The Immortal Emperor* (Cambridge, 1992).

——, *The Reluctant Emperor. A Biography of John Cantacuzene, Byzantine Emperor and Monk, c.1295–1383* (Cambridge, 1996).

Oikonomides, N. "Contribution à l' étude de la *pronoia* au XIIIe siècle," *REB* 22 (1964), 158–175.

——, *Les listes de préséance byzantines des IXe et Xe siècles* (Paris 1972).

——, "Notes sur un praktikon de pronoiaire (juin 1323)," *TM* 5 (1973), 335–346.

——, "La decomposition de l' empire Byzantin à la veille de 1204 et les origins de l' Empire de Nicée: À propos de la *Partitio Romaniae*," in Actes du *XVe Congrès des études byzantines. Rapports et co-rapports*, (Athens, 1976), I, 1–28. reprinted in N. Oikonomides, *Byzantium from the Ninth Century to the Fourth Crusade. Studies, Texts, Monuments* (Aldershot, 1992), Study XX.

——, "L' evolution de l' organisation administrative de l' empire byzantin," *TM* 6 (1976), 125–152.

——, *Hommes d' affairs Grecs et Latins à Constantinople. XIIe–XVe siècles* (Montreal, 1979).

——, "The properties of the Deblitzenoi in the Fourteenth and Fifteenth Centuries," in A. Laiou (ed.), *Charanis Studies: Essays in Honor of Peter Charanis* (New Brunswick, 1980), 176–198.

——, "À propos des armées des premières Paléologues et des compagnies de soldats," *TM* 8 (1981), 353–371.

——, *Byzantine Lead Seals* (Washington, 1985).

——, *A Collection of Dated Byzantine Lead Seals* (Washington, 1986).

——, "Ottoman Influence on Late Byzantine Fiscal Practice," *SüdostF* 45 (1986), 1–24.

——, "The Role of the Byzantine State in the Economy," in A. Laiou (ed.), *The Economic History of Byzantium* (Washington, 2002), 973–1058.

Ostrogosky, G. "Byzance état tributaire de l' Empire turc," *ZRVI* 5 (1958), 49–58.

Ousterhout, R. *Master Builders of Byzantium* (Philadelphia, 2008).

——, "Byzantium between East and West and the Origins of Heraldry," in C. Hurihane (ed.), *Byzantine Art. Recent Studies* (Tempe, 2009), 153–170.

Ousterhout, R. - Bakirtzis, Ch. *The Byzantine monuments of the Evros/Meriç River Valley* (Thessalonica, 2007).

Özgüven, B. "Fortress of Kilid-ül-Bahir," in S. Ćurčić-E. Hadjitryphonos (eds.), *Secular Medieval Architecture in the Balkans, 1300–1500, and its Preservation* (Thessalonica, 1997), 139–143.

Panagiotides, K. "Η επιμέλεια του στρατού κατά τους ύστερους Βυζαντινούς χρόνους (1204–1453)," [The Supplying of the Army in the Late Byzantine Period (1204–1453)] *Byzantina* 24 (2004) 289–313.

Papadopoulos, A. *Versuch einer Genealogie der Palaiologen. 1259–1453* (Munich, 1938).

Petrovič, D. "Firearms in the Balkans on the eve and after the Ottoman Conquests of the Fourteenth and Fifteenth Centuries," in V.J. Parry - M.E. Yapp (eds.), *War, Technology and Society in the Middle East* (London, 1975), 164–194.

Polemis, D. *The Doukai. A Contribution to Byzantine Prosopography* (London, 1968).

Polemis I. *Ο λόγιος Νικόλαος Λαμπηνός και το εγκώμιον αυτού εις τον αυτοκράτορα Ανδρόνικον Παλαιολόγον [The Scholar Nikolaos Lampenos and his encomium to the emperor Andronikos II]* (Athens, 1992).

Popovič, M. "Les forteresses dans les régions des conflits byzantinoserbes au XIVe siècle," in N. Oikonomides (ed.), *Byzantium and Serbia in the 14th Century* (Athens, 1996), 67–87.

——, "The Fortress of Belgrade," in S. Ćurčić - E. Hadjitryphonos (eds.), *Secular Medieval Architecture in the Balkans, 1300–1500, and its Preservation* (Thessalonica, 1997), 128–131.

——, *Beogradska tvrdjava* [*The Fortress of Belgrade*] (Belgrade, 2006) 2nd edition. Book written in Serbo-Croatian with a summary in English.

Powers, J.F. *A Society Organized for War. The Iberian Municipal Militias in the Central Middle Ages. 1000–1284* (Berkeley, 1988).
Ramsay, W.M. *The Historical Geography of Asia Minor* (London, 1890 repr. New York, 1972).
Raybaud, L.-P. *Le gouvernment et l' administration central de l' empire byzantine sous le premières Paléologues* (Paris, 1968).
Redford, S. "A Grammar of Rum Seljuk Ornament," *Mesogeios* 25/26 (2005), 283–310.
Rogers, R. *Latin Siege Warfare in the Twelfth Century* (Oxford, 1992).
Sachedina, A. "Al-Khums: The Fifth in the Imami Shi'I Legal System," *Journal of Near Eastern Studies* 39 (1980), 275–289.
Schmitt, O.-J. "Kaiserrede und Zeitgeschichte im späten Byzanz: Ein Panegyrikos Isidoros von Kiew aus dem Jahre 1429," *JÖB* 48 (1998), 209–242.
Schneider, J. "Sir Nicole Louve: citain du Metz," in P. Contamine (ed.), *La noblesse au moyen âge* (Paris, 1976).
Schreiner, P. "Zur Geschichte Philadelpheias im 14. Jahrhundert (1293–1390)," *OCP* 35 (1969), 375–431.
Seibt, W. "Das Monogramn-Siegel eines Theodoros Dukas Synadenos aus der frühen Palaiologenzeit," *JÖB* 40 (1970), 271–273.
Setton, K. *Catalan Domination of Athens. 1311–1388* (Cambridge, 1948).
Settia, A. *De re militari: pratica e teoria nella guerra medieval* (Rome, 2008).
Sinor, D. "The Inner Asian Warriors," *Journal of the American Oriental Society*, 101 (1981), 133–144.
Smyrlis, K. "The State, the Land, and Private Property: Confiscating Monastic and Church Properties in the Palaiologan Period," in D. Angelov (ed.), *Church and Society in Late Byzantium* (Kalamazoo, 2009), 58–87.
Ševčenko, I. *Études sur la polémique entre Théodore Métochite et Nicéphore Choumnos* (Brussels, 1962).
——, "Theodore Metochites, the Chora, and the Intellectual Trends of His Time," in P.A. Underwood (ed.), *The Kariye Djami. Studies in the Art of the Kariye Djami and its Intellectual Background*, 4 vols. (Princeton, 1975), IV, 19–84.
Shawcross, T. *The Chronicle of Morea. Historiography in Crusader Greece* (Oxford, 2009).
Soulis, G.C. *The Serbs and Byzantium during the reign of Tsar Stephen Dušan. 1341–1355 and his Successors* (Athens, 1995).
Stavridou-Zafraka A. *Νίκαια και Ήπειρος τον 13ο αιώνα. Ιδεολογική αντιπαράθεση στην προσπάθεια τους να ανακτήσουν την αυτοκρατορία* [*Nicaea and Epiros in the Thirteenth Century. Ideological Conflict in their Attempt to Recover the Empire*] (Thessalonica, 1991).
Stouraitis, I. *Krieg und Frieden in der politischen und ideologischen Wahrnehmug in Byzanz (7.–11. Jahrhundert)* (Vienna, 2009).
Sullivan, D. "Tenth-Century Byzantine Offensive Siege Warfare: Instructional Prescriptions and Historical Practice," in N. Oikonomides (ed.) *Byzantium at War* (Athens, 1997), 179–200.
Sygkelou, E. *Ο πόλεμος στον δυτικό ελλαδικό χώρο τον ύστερο μεσαίωνα* [*Warfare in Western Greece in the Late Middle Ages*] (Athens, 2008).
Svoronos, N. "Le serment de fidélité a l' empereur byzantine et sa signification Constitutionnelle," *REB* 9 (1951), 106–142.
Tourta, A. "Fortifications of Gynaikokastron," in S. Ćurčić, – E. Hadjitryphonos (eds.), *Secular Medieval Architecture in the Balkans, 1300–1500, and its Preservation* (Thessalonica, 1997), 110–113.
Trapp, E. "Beiträge zur Genealogie der Asanen in Byzanz," *JÖB* 25 (1976), 163–177.
——, *Prosopographisches Lexikon der Palaiologenzeit* (Vienna, 1976–1996).
——, *Lexikon zur byzantinischen Gräzität* (Vienna, 1996–2001).
Troianos, S. "Kastroktisia," *Byzantina* 1 (1969), 39–57.

Tsouris, K. - Mprikas, A. *Το φρούριον του Πυθίου και το έργο της αποκατάστασης του. Προκαταρκτική Ανακοίνωση* [*The Fortress of Pythion and the Work of its Restoration. Preliminary Report*] (Kavala, 2002).

Typaldos, G. "Είχον οι Βυζαντινοί οικόσημα;" [Did the Byzantines have Heraldic Emblems?] *Epeteris Etairias Vyzantinon Spoudon* 3 (1926), 206–222.

Underwood, P.A. (ed.), *The Kariye Djami*, 4 vols. (Princeton, 1966–1975).

Van Milingen, A. *Byzantine Constantinople* (London, 1899).

——, *Byzantine Churches in Constantinople. Their History and Architecture* (London, 1912).

Varanini, G.K. "Mercenari tedeschi in Italia nel Trecento: problema e line di ricerca," in S. De Rachewiltz - J. Riedman (eds.), *Kommunikation und Mobilität im Mittelalter* (Vienna, 1995), 159–178.

Vásáry, I. *Cumans and Tartars. Oriental Military in the Pre-Ottoman Balkans, 1185–1365* (Budapest, 2005).

Verbruggen, J.F. *The Art of Warfare in Western Europe during the Middle Ages*, tr. S. Willard - S.C.M. Southern (Suffolk, 1997).

Verpeaux, J. *Nicéphore Choumnos, homme d'état et humaniste byzantin (ca. 1250/1255–1327)* (Paris, 1959).

——, "Les *Oikeioi*. Notes sur l'histoire institutionelle et sociale byzantin," *REB* 23 (1965), 89–99.

Vryonis, S. *The Decline of Medieval Hellenism in Asia Minor and the Process of Islamization from the Eleventh through the Fifteenth Century* (Berkeley, 1971).

——, "Byzantine and Turkish Societies and their Sources of Manpower," in V. Parry (ed.), *War Technology and Society in the Middle East* (London, 1975).

Waley, D. "The Army of the Florentine Republic from the Twelfth to the Fourteenth Century," in N. Rubinstein (ed.), *Florentine Studies. Politics and Society in Renaissance Florence* (London, 1968), 70–108.

——, "Condotte and Condottieri in the Thirteenth Century," *Proceedings of the British Academy*, 61 (1975), 337–371.

Weiss, G. *Joannes Kantakuzenos—Aristokrat, Staatsmann, Kaiser und Mönch—in der Gesellschaftsentwicklung von Byzanz im 14. Jahrhundert* (Wiesbaden, 1969).

Wolfe, M. "Siege Warfare and the *Bonnes Villes* of France during the Hundred Years War," in I.A. Corfis - M. Wolfe (eds.), *The Medieval City under Siege* (Woodbridge, 1995), 49–66.

Woodhouse, C.M. *George Plethon Gemistos: The Last of the Hellenes* (Oxford, 1986).

Zachariadou, E. "The Conquest of Adrianople by the Turks," *Studii Veneziani* 12 (1970), 211–217.

——, "Manuel II Palaeologus on the Strife between Bayezid I and Kadi Burhan al-Din Ahmad," *Bulletin of the School of Oriental and African Studies*, 43 (1980), 471–481.

——, *Trade and Crusade: Venetian Crete and the Emirates of Menteshe and Aydin. 1300–1415* (Venice, 1983).

——, *Ιστορία και θρύλοι των παλαιών Σουλτάνων (1300–1400)* [*History and Legends of the old Sultans*] (Athens, 1991).

Zafeiris, K. "The Issue of the Authorsip of the Synopsis Chronike and Theodore Skoutariotes," *REB* 69 (2011), forthcoming.

Zakythinos, D. *Crise monétaire et crise économique à Byzance du XIIIe–XVe siècle* (Athens, 1948).

——, *Le despotat grec de Morée*, 2 vols. (Paris, 1932/1953 repr. London, 1975).

Živojinović, M. "La frontier serbobyzantine dans les premières décennies du XIVe siècle," in N. Oikonomides (ed.), *Byzantium and Serbia in the 14th Century* (Athens, 1996), 57–66.

INDEX

Acciaiuioli, Angelo, Cardinal, 140
Acciaiuioli family, 103, 227
Acciaiuioli, Nerio, duke of Athens, 103, 140
Achaia, principality of, 20, 28, 101, 103, 140, 188, 199, 202, 203, 210, 227
Acheloos, battle of (1359), 112
Achilleid, 50, 66, 84
Achyraitai, group of soldiers, 70, 109
adnoumion, 156
Adrianople (Edirne), 65, 93, 138, 165, 168, 189
Adrianople, battle of (1205), 105, 198, 201
Adriatic, 193
Aegean, 1, 2, 29, 34, 37, 69, 107, 110
Aetos, fort, 188
Ainos (Enez), 36 n. 87, 116
Akıncı, 215
Akova, 28
Akropolites, George, 5, 6, 19, 47, 49, 62, 67, 82, 86, 94, 105, 114, 138, 146, 152, 154, 171–173, 198, 199, 218
Alans, 104, 127, 179
Alan mercenaries, 120, 121, 123, 125, 126, 129, 134, 135, 217
Albania, Albanians, 2, 32, 50, 63, 89, 101, 133, 138, 140, 144, 146, 150–152, 198, 202, 218
Albanian mercenaries, 103, 106, 112
Alexander, Byzantine emperor (912–913), 23
Alexios I Komnenos, Byzantine emperor (1081–118), 118, 213
Allagia, 8, 82–93
Almugavars, 101, 121, 212
Amadeo of Savoy, 56, 212, 213
Amphipolis, 159
Anastasioupolis, 159
Anatolikon, tower, 188
Andronikos II Palaiologos, Byzantine emperor (1282–1328), 2, 5, 6, 13, 14, 15, 25–31, 37, 46, 66, 72, 76, 83, 86, 88, 107, 108, 120, 122, 125, 128, 130, 135, 137, 141, 143, 144, 146, 157–160, 166, 167, 226
Andronikos III Palaiologos, Byzantine emperor (1328–1341), 2, 5, 30–35, 38–40, 45, 46, 54–57, 69, 72, 78, 80, 85, 92, 93, 102, 108, 111, 113, 115, 130, 137, 143, 146, 151, 154, 159, 161, 165, 178–180, 184, 185, 204, 205
Andronikos IV Palaiologos, Byzantine emperor (1376–1379), 38
Androusa, 137
Angelokastron, 50, 150
Angelos, Doukas Demetrios, despot in Thessalonica (1244–1246), 145
Angelos John, nephew of John VI, 35, 61
Angelos John, Nicaean general, 82, 166
Angelos John Doukas, military commander in Asia Minor, 24, 30
Angelos Theodore, bastard son of Michael II Doukas Angelos, 152
Angevins, 121
Angold M., 88
Anhialos (Pomorie), 37
Anna of Savoy, 35, 54
Antikensorai, 155
Antioch (Antakya), 53 n. 37
Antioch-on-the Meander, battle of (1211), 17, 18, 49, 135, 212
Aplekton (*pezeuma, pessimon*), 145
Apokaukos Alexios, 33–35, 39, 47, 72, 161
Apros, battle of (1305), 79, 96, 166, 214
Apros, 178, 179, 184, 217
Arabs, 17, 148, 208
Arcadius, emperor (395–408), 129
Argyrokastron (Gjirokastër), 106
Arkadioupolis (*Lüleburgaz*), 36, 159
Argos, 106
Arsenios, patriarch, 80 n. 21
Arsenite movement, 80
Arta, 32, 181, 182
artillery weapons (see also bombards, cannon, gunpowder, siege engines), 185, 194
Asan, Andronikos, Byzantine general in the Peloponnese, 28, 85 n. 44
Asan, Constantine, Byzantine military commander, 78
Asan, Manuel, head of the imperial *allagion*, 85, 219

asapi, Venetian troops, 150
Asia Minor, 1, 2, 8, 10, 13, 16, 21–28, 30, 31, 37, 62, 69, 70, 77–79, 81, 84, 86–88, 94, 95, 107, 111, 120, 124, 137, 145, 152, 153, 157–159, 162, 163, 166, 168, 175, 176, 199, 207, 209, 210, 221, 223, 226
Aşık Paşazade, 177, 179, 189
Athanasios I, Patriarch of Constantinople, 162, 166
Athos, 90, 163
Attaleiates, Michael, 86
Attica, 22, 72
Autoreianos, Michael, Patriarch of Nicaea, 16–17
Avars, 206
Axios river, 72, 103, 139
Aydın, Turcoman principality, 32, 180, 212
azabs, Ottoman troops, 150

Baggage train, 142, 143
Bapheus, battle of (1302), 120, 121
Bardas, Phokas, Byzantine general, 62
Bartusis M., 11, 88, 89, 98, 139, 168, 187
Basil I, Byzantine emperor (867–886), 23
Basilika, 147
Bayezid I, Ottoman ruler (1389–1402), 10, 43, 190
Belgrade, 191, 192, 194
Bellagrada (Berat), 102, 141, 198, 201, 202, 211, 216, 218
Belting, H., 58
Berenguer d' Entença, Catalan commander, 124
Bertradon de la Brocquière, Burgundian traveler, 55, 56
Bessarion, Cardinal, 95, 98, 196
Bithynia, 24, 31, 44, 107, 158, 178
Black Sea, 37, 158, 183
Blemmydes, Nikephoros, 14, 56, 61, 119
Bohemond, 118
Bombards (see also artillery weapons, cannons, gunpowder), 188–190
Booty (See also plunder, spoils of war), 67, 71, 81, 93, 94, 122, 142, 146–150, 153, 164, 165, 168, 175, 214, 218, 221, 226
Bouchard, William, French knight, 54
Boucicaut Marshall, Jean le Meingre, French general, 56, 212, 213
Branković, George, Serbian despot, 162, 194
Brienne, Norman general, 118
Bruidy, John, French mercenary of John VI, 59
Bulgaria, Bulgarians, 21, 22, 30, 31, 36, 41, 46, 49, 51, 64, 67, 85, 104, 105, 111, 115, 142, 144, 151, 152, 164, 171, 172, 176, 179, 198, 199, 201, 205, 213, 218, 227
Burhan ed-din Ahmed, 43
Byzantine archers, 210
Byzantine cavalry, 91, 199, 209, 211
Byzantine companies of soldiers, 108
Byzantine fleet, 29, 106, 107, 141, 226
Byzantine mercenaries, 104

Canon law, 16
Cannon (see also bombards, artillery weapons, gunpowder), 16, 170, 190–195
Caria, 24
Carlo I, Tocco, duke of Cephalonia, 7, 150, 188
Catalan duchy of Athens, 103
Catalan Grand Company, 8, 70 n. 111, 72, 80, 96, 101, 109, 111, 112, 121–129, 134, 142, 148, 166–168, 169, 186, 187, 200, 210, 212, 220, 223
Catalan mercenaries, 104, 109, 131
Catanea, Andreolo, Genoese ruler of Phokaia, 32
Catanea, Domenico, Genoese ruler of Phokaia, 180
Cayster river, 24, 27
Cavalry/mounted archers, 55, 64, 67
Cephalonia, 7, 188
Chadrenos, Michael, Byzantine general, 167
Chalkidike, 109
Chalkokondyles Laonikos, 7, 189, 190, 192, 215
Charioros, 171
Charles I of Anjou, King of Sicily, 26, 102, 141, 198, 210
Charles VI, King of France, 213
Charsianon *theme*, 84
Chevedden, P., 174
Chilandar monastery, 163
Chios, 32, 34, 39, 59, 115, 161, 180
chivalric literature, 68
Choirovoskos, Byzantine commander, brigand, 69, 210
Choniates, Niketas, 17, 18, 20, 49, 201
Chortiates, battle of (1383), 82, 116, 172
Choumnos, Nikephoros, *megas logothetes*, 26, 72

Choumnos, military commander in Adrianople, 168
Christoupolis (Kavala), 35, 36, 92
Chronicle of the Morea, 7, 22, 52, 54, 67, 84, 110, 112, 113, 140, 186, 201
Chronicle of the Toccos, 7, 50, 98, 150, 188, 189
Chryseia Gate (Constantinople), 95
Chrysobull, 15, 35, 36, 83, 122, 143, 144, 160
Chrysoloras, Manuel, 159
Clarenza, 188
Clavijo Ruy González de, Castilian diplomat, 190
comes stabuli, 117
Companies of soldiers, 108–111, 117
Conestabilia, conestabileria, 117
Constantine VII, Byzantine emperor (912–959), 114, 161, 167
Constantine XI Palaiologos, Byzantine emperor (1448–1453), 7, 95, 98, 137
Constantinople, 1, 8, 16, 19, 20, 22–23, 35, 37, 38, 56, 82, 93–95, 102, 108, 113, 115, 129, 137–139, 160, 162, 163, 165–167, 171, 173, 176, 179, 182, 183, 190–194, 204, 209, 213–215, 219, 221, 224
Corinth, 106
Cretan imperial guards, 115
Cretan mercenaries, 107, 120, 125, 126, 135
Crossbow, 63, 110, 174, 178, 179, 189, 190
Crusades, Crusaders, 1, 4, 17, 37, 186, 211
Cuman mercenaries, 84, 104–106, 112, 115, 116, 135, 199–203, 217, 224
Cumans, 75, 102, 198, 199, 201

Dakivytza, 204
Dennis, G., 9
Devlitzenoi, 82
Devlitzenos Demetrios, soldier of the *mega allagion* of Thessalonica, 91
Di Bartolomeo Michelozzo, Florentine architect and engineer, 194
Didymoteichon, 36, 86, 93, 130, 160, 165, 168, 183, 189, 199
dikavalloi, 89, 90
Dobrotič, 67
Docheiariou monastery, 90
Doukas, 7, 99, 115, 152, 190, 192
Doukas, Manuel Andrianos, soldier of *mega allagion* of Thessalonica, 91
Doukas, Nostongos, Byzantine general, 177
Dragamesto, fortress, 188
Dragotas, head of Menik, 82, 83
Dubrovnik, 194
Dyrrachion (Durres), 119

Edessa (Vodena), 71, 102, 144, 184
Eirene Asanina Kantakouzene, wife of John VI, 102
Eisagoge, 23
England, 146
Enveri, 39
Epiros, 1, 2, 8, 17, 20, 21, 22, 27, 32, 33, 39, 40, 63, 81, 92, 101, 102, 105, 111, 137, 146, 151, 165, 185, 189, 198, 200, 202, 210, 227
epi tou stratou, 154, 155
Epivatai, 161
Eressos, 180
Euboia, 22
Eustathios, bishop of Thessalonica, 173
Evrenos, Ottoman frontier lord, 43
Exisoseis, 78

Firearms (see also artillery weapons, bombards, cannons, gunpowder), 190, 194
Foss C., 158, 192
Fourth Crusade, 1, 16, 22, 32, 107
France, *Frankia*, 22, 117, 146, 213
Frankish Greece, 50, 54, 59, 103, 110, 117, 133, 139, 150, 188, 196, 221, 224
Frankish heavy cavalry, 63, 201, 203, 217
Frankish mercenaries, 108, 118
Franks, 23
Frederick II Hohenstaufen, Western Emperor (1194–1250), 19
Frederick III of Aragon, king of Sicily, 126
French knights, 102

Galata, 109, 173, 190, 191, 209
Galikos river, 102
Gallipoli (Gelibolu), 41, 43, 45, 196, 212
gasmouloi, 96, 107
Gattilusio, Francesco, Genoese ruler of Lesvos, 116
Gattilusio, Nicolo, Genoese ruler of Ainos, 116
Gavrielopoulos Michael, semi-independent ruler of Thessaly, 160, 167
Gavrielopoulos Stephen, semi-independent ruler of Thessaly, 181

Genoa, Genoese, 109, 131, 190, 191, 213, 219, 227
Georgia, 105
Germanos, patriarch, 20
Germans, 200
German knights, 102, 103
German mercenaries, 101, 102, 112, 113, 130–132, 152, 179
Germiyan, Turcoman principality, 162
Giustiniani, Giovanni, Genoese defender of Constantinople (1453), 192
Glabas, Michael Doukas Tarchaneiotes, Byzantine general, 48, 49, 58, 60, 63, 69
Golden Gate (Constantinople), 132, 187, 190
Goutziokostas A., 151
Grant, John, engineer defender of Constantinople (1453), 102
Gratianou (Gratini), 153
Gregoras Nikephoros, 6, 8, 28, 30, 35, 49, 50, 55, 57, 69–73, 81, 95, 96, 104, 106, 120, 125–128, 135, 164, 165, 167, 176, 198, 200, 201, 204, 205, 207, 212, 214, 217, 218
Gregory of Cyprus, patriarch, 20–22, 26, 27
gunpowder, (see also artillery weapons, bombards, cannons), 187–196, 224
Gynaikokastron, fortress, 102, 139, 158, 163

Hacı İlbey, Ottoman frontier lord, 189
Haldon, J. 84, 150
Halil, Turcoman adventurer, 72, 73, 112, 217
handgunners, 192, 202
Harmenopoulos, Constantine, 147
Hawkins, E.J.M., 58
Hawkwood, John, English condottiere, 134
helepolis, 170–173, 180
Henry of Flanders, Latin emperor of Constantinople, (1206–1216), 17, 135, 166, 201
Herakleia on the Black Sea, 24, 176
Heraldry, 57–59, 222
Hexabiblos, 147
Hexamilion, 95, 106, 159, 160, 191
Hippodrome (Constantinople), 55
Holobolos, Manuel, 20
Holy War, 221
Honorius, Western emperor (395–423), 161
Hospitallers, 103
Hugue Rousseau of Sully, general of Charles I of Anjou in Albania, 202
Hungarians, 105, 114, 191, 194
Hyperpyra, 76, 144

Iagaris, Manuel, 162
Ignatius of Smolensk, Russian traveler, 187
Innocent III, Pope, 17 n. 17, 201
Ioannina, 83, 160
iovola, 174
Isidore, bishop of Kiev, 37
Italian authors, 134
Italy, 4, 68, 111, 113, 117–119, 122, 134, 200
Ivan the Russian, commander in Philippoupolis, 104
Izz al-Din Kayka' us II, Seljuk Sultan, 214

Jacob, bishop of Bulgaria, 19
Jacques of Baux, claimant of the principality of Achaia, 104
James, Latin bishop of Argos, 140
Janissary troops, 214
Jean de la Roche, ruler of Athens, 217
Joanna of Naples, Angevin claimant to the throne of Albania, 103
John Alexander, Bulgarian emperor (1331–1371), 46, 49, 185
John II Komnenos, Byzantine emperor (1118–1143), 172
John II Palaiologos, marquis of Monferrat, 134
John III Asen, Bulgarian emperor, (1279–1280), 28 n. 60, 60, 85 n. 44
John III Vatatzes, Nicaean emperor (1221–1254), 1, 18, 19, 77, 104, 119, 138, 142, 145, 154, 166, 171–173, 209
John IV Laskaris, Nicaean emperor (1258–1261), 80 n. 21
John V Palaiologos, Byzantine emperor (1341–1391), 35, 36, 37, 45, 46, 113, 131, 182, 183, 212, 213, 219
John VI Kantakouzenos, Byzantine emperor (1347–1354), 2, 4, 5, 9, 30–36, 38–40, 45–52, 54, 56, 59, 61–66, 69, 71, 72, 78–80, 86, 89, 91, 93, 96, 101–104, 110, 112, 115, 130–132, 135, 137, 139, 141, 142, 144–148, 151–154, 159, 161, 164, 165, 167, 168, 170, 176–179, 181–186, 204–206, 214, 216, 218, 219, 222, 225

John VII Palaiologos, Byzantine emperor (1390–1391), 38, 187
John VIII Palaiolgos, Byzantine emperor (1425–1448), 7, 37, 38, 188
Joseph the Philosopher, 15
joustra, 54, 57
Juan de Spinia, Latin commander in Phokaia, 181
Just Wars, 16, 23, 40, 221

Kaikhusraw I, Seljuk Sultan (1205–1211), 18
Kalekas, John, patriarch, 35
Kallimachos and Chrysorrhoe, 51, 66
Kaloeidas, Manuel, soldier of the *vassilikon allagion*, 86, 91
Kananos, John, 168, 171, 190
Kanina, 144, 218
Kantakouzene Theodora, daughter of JohnVI, wife of Orhan, 40
Kantakouzenos, Manuel, Byzantine despot of the Morea (1349–1380), 36, 112
Kantakouzenos, Matthew, son of John VI, 36, 218
Kantakouzenos Palaiologos, George, 194
Karamehmed, Turcoman adventurer, 40
Kariye Cami (Chora Monastery), 60
Karytaina, 28
Kassianos, Byzantine military commander in Mesothynia, 29, 30
kastroktisia, 159, 160
kastrophylax, 169
Katharos, Michael, bastard son of the despot Constantine Palaiologos, 31
Katoche, tower, 189
kavallarioi, 59, 108, 130, 132, 166
kephale, 33, 83, 143, 169
Kilid-ül-Bahir, fortress, 196
Kinnamos, John, 172
Klazomenitai, group of soldiers, 90 n. 60, 109
Kolias, G., 187
Komnene, Anna, 118
Komnenos, Adrianos, 118 n. 94
Komnenian period, 4, 60
Kontophre, Byzantine governor of Mesothynia, 62
korte, 114
kortinarioi, 113, 114
Kosmidion, 138 n. 2
Koteanitzes, Leo, 76
Kotertzes, Nikolaos, Nicaean general, 62
Koumoutzena (Komotini), 145, 154
kouropalates, 169
koursatoroi, 150
kourseuontes, 149, 150
koursores, 149
krites tou phossatou/stratopedou, 150, 151
Kritoboulos from Imbros, 170
Krytzimos, 171
Kütahya, 162
Kydones, Demetrios, *mesazon*, 9, 40–43, 45, 116, 117
Kyzikos, 123

ladders (*klimakes*), 179, 183–185, 191
Ladolfo, Matteo, captain of Carlo I Tocco, 188
Laiou A., 23
laisai, 175
Lampenos, Nikolaos, 27, 77
Lantzaretos, armed servant of Kantakouzenos, 154
Laskarids, 16, 18, 24, 106
Laskaris, George, 166
Laskaris, Manuel, 67, 85, 86
Laskaris, Michael, 144
Latins, 4, 39, 107, 137, 200, 221
Latin empire of Constantinople, 1, 17, 105, 138, 171, 172, 198, 201, 214
Latin mercenaries, 17, 18, 19, 57, 84, 104, 113, 115–119, 130, 179, 212
Lazarević, Stefan, Serbian prince, 194
Lentiana, 171
Leo the Deacon, 53
Leo VI the wise, Byzantine emperor (886–912), 23
Leonard, Latin bishop of Chios, 102, 192
Leontaris Bryennions, Manuel, 162
Lesvos, 32, 39, 116, 180, 213
Lips (Fenari Isa Cami), 58
Limpadarios, 138 n. 2
Louis of Evreux, count of Beaumunt-le-Roger, 103
logothetes tou stratiotikou, 156
Lydia, 27, 158

Macedonia, 2, 21, 29, 36, 43, 62, 76, 84, 92, 93, 104, 105, 111, 128, 130, 137, 140, 142, 146, 158, 167, 183, 218, 220
Madytos, 124, 186
Magistros, Thomas 14, 15, 64, 77, 128, 129, 135, 146, 220
Magnesia, 126, 137
Makrenos, Byzantine general in the Peloponnese, 140, 216
Makrolivada, 138

Mallet M., 68
Manfred, Hohenstaufen king of Sicily, 198
mangana, 174, 175
Mango C., 58
Mantzikert, battle of (1071), 86
Manuel I Komnenos, Byzantine emperor (1143–1180), 53, 56, 114, 172
Manuel II Palaiologos, Byzantine emperor (1391–1425), 7, 9, 10, 38, 41–43, 61, 63, 65, 78, 82, 99, 106, 116, 117, 133, 159, 189, 213, 214, 227
Manuelites Nikolaos, Byzantine governor of Nicaea, 62
Margarites, Constantine, 85, 86, 199
Maritza, battle of (1371), 37, 79
Maroules, Nikolaos, *pronoia* holder, 90 n. 60
markets, 145–146
Martuman, Turcoman adventurer, 40
Matthew, bishop of Ephesos, 29, 47
Mazaris' jurney to Hades, 189
McGeer E., 175
Meander river, 24, 25, 104, 158
megaloallagitai, 89, 90
Megara, 103
megas admouniastes, 156
megas domestikos, 46, 49, 147, 148, 155, 156
megas drouggarios tes vigles, 155, 156
megas konostaulos, 48, 117, 118
megas stratopedarches, 47, 140, 141
megas tzaousios, 85
Mehmed I, Ottoman ruler (1413–1421), 191
Mehmed II, Ottoman ruler (1451–1481), 196
Melanas river, 78
Melnik, 82, 116, 145, 183
Menander Rhetor, 15
mensores, 155
Menteşe, Turcoman principality, 26, 176, 178
Mesemvria (Nezibar), 36, 169, 213
Mesene, 154
Mesothynia, 29, 31, 62, 84, 92, 204
Metochites Theodore, *megas logothetes*, 13–14, 21, 25, 26, 129, 157–159
Michael II Doukas Angelos, ruler of Epiros (1230–1271), 22, 25 n. 49, 92 n. 66, 146, 198
Michael VIII Palaiologos, Byzantine emperor (1259–1282), 1, 2, 6, 16, 20–26, 29, 30, 32, 48, 50, 61, 62, 69, 70, 76, 77, 83, 95, 112, 118, 139, 143, 144, 152, 158, 160, 161, 167, 171, 173, 174, 202, 203, 207, 210, 226
Michael IX Palaiologos, Byzantine co-emperor (1294–1320), 14, 31, 45–47, 77, 79, 107, 108, 114, 119, 126, 137, 166, 217
Michael Šišman, Bulgarian emperor, (1323–1330), 31, 46, 151
military manuals/ treatises, 4, 63–65, 84, 94, 158
minsouratores, 155
Mirrors of princes, 14
mitaton, 143–144, 153, 223
Momčilo, brigand, 145, 154, 219
Monemvasia, 186, 203
Mongols, 67, 145, 176, 177
monograms, 59–60
monokavalloi, 89, 90
Monomachos, Senacherem, Byzantine general, 181
Moreas, 2, 7, 22, 23, 37, 41, 43, 95, 96, 106, 112, 116, 133, 215
Mourike, D. 58
Mourtatoi, 113, 114
Mouzalon brothers, 47, 48
Mouzalon, Leo, Byzantine general, 107, 120, 208
Muntaner Ramon, 7, 8, 121–124, 142, 148, 186, 212
Murad I, Ottoman ruler (1362–1389), 37, 41
Murad II, Ottoman ruler (1421–1451), 37, 115, 171, 177, 190–192
Mylassa, 84, 92
Mythimna, 180
Mytilene, 124, 180, 181

Navarrese and Gascon Companies of soldiers, 41, 98, 99, 103, 106, 140
Neokastra, 84, 85, 92
Neopatras, 50, 116
Nestogos, Theodore, Nicaean general, 82
Nicaean army/archers, 173, 174, 209–210
Nicaea, city of, 25, 62, 178, 208
Nicaea, empire of, 1, 5, 6, 17–20, 56, 88, 95, 105, 152, 153, 163, 201, 214
Nicaean frontier guards, 93–95
Nikephoros II Doukas Angelos, despot of Epiros (1335–1340), 33, 112
Nikephoros II Phokas, Byzantine emperor (963–969), 15, 17, 53, 56, 62
Niketiatou, 204

Nikomedia, 158, 176, 203, 208
Nikopolis, battle of (1396), 213
Niphon, bishop of Kyzikos, 162
Normans, 118, 173
Notaras, Loukas, 192
Nymphaion, 161

Ochrid, 164
Oikonomides, N., 11, 59, 88, 89
Olbofaci Andreas and Theodore, *asapi* of Venice, 150
On Skirmishing, 71
On Strategy, 156
Orhan, Ottoman ruler (1326–1362), 28, 31, 40, 165, 178, 204
Osman, Ottoman ruler (1288–1326), 28, 29, 208
Ostrovos, 142
Ottomans, 2, 7, 31, 37–39, 40, 44, 61, 79, 80, 82, 85, 95, 97, 102, 106, 115, 120, 133, 159, 168, 176–178, 185, 189–191, 195, 196, 204–206, 212–215, 221, 227
Ottoman army, 7, 36, 43, 150, 151, 215
Oumbertopoulos, Constantine, 118 n. 94
Oumbertopoulos, *kouropalates*, 169
Oumbertopoulos, *megas tzaousios*, 169
Ouranos, Nikephoros, 94

Pachymeres, George, 5, 6, 8, 22–24, 26, 28, 62, 69, 70, 76, 77, 79–81, 84, 87, 91–93, 95, 106, 108, 116, 120–123, 125–127, 134, 135, 141, 142, 144, 153, 167–169, 173, 174, 177, 200, 205, 208–210, 217
palace guard units, 113–115
Palaiologina, Eirene Raoulaina, 60
Palaiologos, Andronikos, author of *Kallimachos and Chrysorhoe*, 51 n. 26
Palaiologos, Andronikos, (father of Michael VIII), 62
Palaiologos, Constantine, (brother of Michael VIII), 62, 199
Palaiologos, Constantine, *porphyrogennetos*, (son of Michael VIII), 24, 28, 50, 62, 76
Palaiologos, Constantine, despot (son of Andronikos II), 31
Palaiologos, John, despot (brother of Michael VIII), 21, 24, 60, 62, 69, 70, 87, 92, 105, 114, 140, 199, 201, 217, 227
Palaiologos, John, caesar, (son of Constantine *porphyrogennetos*), 50
Palaiologos, Michael (son Andronikos III), 113
Palaiologos, Theodore, marquis of Montferrat (son of Andronikos II), 4, 68, 134
Pammakaristos monastery (Fethiye cami), 49, 58
Pamphylon, 168, 183
Panagiotidis, K., 11
Panaretos, John, soldier of the *mega allagion* of Thessalonica, 90 n. 60
Panaretos, Martinos, soldier of the *mega allagion* of Thessalonica, 90 n. 60
Paphlagonia, 81, 84, 92, 207, 210
Paramonai, 113, 114
Paroikoi, 166
Patras, 98, 137, 189
Patrikiotes, 79
Pazarlı, brother of Orhan, 205
Pege Gate (Constantinople), 162
Pelagonia, battle of (1259), 22, 65, 113, 116, 199, 200, 202, 211, 217
Pelekanos-Philokrene, battle of (1329), 31, 80, 85, 110, 203–205, 207, 214
Peloponnese, 1, 8, 21, 22, 95, 96, 98, 99, 101, 102, 106, 108, 112, 133, 134, 137, 153, 159, 216
pendamerea, 150
pentamoiria, 147, 150
Peralta, Juan, commander of Catalan mercenaries, 109
Peristitza, 171
Peritheorion, 159, 183, 185
Phanarion, 160
Phakrases, 67
Philadelphia, 25, 29, 37, 47, 69, 70, 94, 162, 209
Philanthropenos, Alexios, Byzantine general in Asia Minor and the Aegean, 28, 30, 69, 70, 107, 120, 125–127, 180
Philes, Manuel, 48, 49, 56, 222
Philes, Palaiologos, Byzantine commander, 72, 73, 217
Philip of Swabia, 105
Philip of Taranta, prince of Achaia, titular emperor of Constantinople, 32
Philippoupolis (Plovdiv), 102, 104, 110, 179, 180, 185, 196
phytilia, 149
phloriatikon, 159, 160
Phokaia, 32, 39, 40, 180–182, 185
Phokas, Bardas, Byzantine general, 62
Phrangopoulos, Constantine, soldier of *vassilikon allagion*, 86, 91
Phrygia, 104

Piccardy, 116
Pinkernes, 69
Planoudes, Maximos, 14, 70
Plato, 124
Pletho, George Gemistos, 96–98, 133, 160
Plunder (see also booty, spoils of war), 137
Poimamenon, battle of (1223/1224), 138, 171, 172
Poujoize, William, French mercenary of John VI, 59
praecepta militaria, 84
Preasnitsa, 76
Preljub, Serbian general, 184
Prilep, 146
primmikerios of the vardariotai, 114
Prinitza, battle of (1262), 203, 211
prisoners, 122, 179
Prodromos monastery, 143
prokoursatores, 149
pronoia, 46, 75–82, 83, 93, 94, 108, 109, 112, 121, 128, 138, 142, 153, 184, 199, 216, 221, 222
proselantai, 93, 96
prostagma, 76, 83, 143, 144
protasekrites, 23
protonotarios, 139
protosevastos, 50
protostrator, 78, 149, 150
Prusa (Bursa), 97, 179
Pseudo-Kodinos, 4, 87, 89, 91, 114, 140, 147–151, 154–156, 212
Pseudo-Lachanas, 69
Pythion, tower, 161, 189, 193

Raidestos (Tekirdağ), 37, 168
Raoul, Alexios, Byzantine general, 62
Raoul, Demetrios, 106
Raoul, John, 62
remission of sins, 16, 18
Rentina, fortress, 163
Retzio, 204
Rhodes, 119
Rimpsas, Nikephoros, Byzantine general, 115, 116
Roger de Flor, head of the Catalan Grand Company, 121–124
Rogoi, 182
Romanos Gate (Constantinople), 190
Rosokastron, battle of (1332), 49, 72, 86, 111, 185
Roupel, 218
Rouphianon monastery, 166
Russia, 105
rymparika, 150

Sakkoi, 154
Sangarios river, 139, 158, 207
Sanudo, Marino, 8
Sardis, 176, 177
Saruhan, Turcoman principality, 32, 165, 180, 212
Saventzes, Michael, 90 n. 60
Schreiner P., 89
Seljuks, 1, 17, 18, 104, 135, 209, 212
Selymvria (Silivri), 37, 154, 166
Senacherem, Michael, Byzantine general, 23, 24
Serbia, Serbians, 2, 13, 27, 30, 34, 37, 40, 41, 62, 64, 72, 82, 102- 104, 110, 112, 119, 130, 131, 152, 162, 164, 199
Serres, 67, 88, 143, 154, 182, 184
Servia, 184
Sevastopoulos, 110, 205, 206
Sgouros, judge of the army, head of a company of soldiers, 109
Sgouros, *stratopedarches ton tzangratoron*, 111
Sguros, Bua, Albanian chieftain, 150
Sicily, 121
Siderokastron, 158
siege engines, 171, 173, 180, 186, 189, 224
siege tower, 179, 180, 185, 196
sipahis, 196
Skamandros river, 70 n. 111
Skopelos, 170
Skoutariotes, Theodore, 6
Slavic tribes, 133, 140
Smallholding soldiers, 11, 104
Smederevo, 194
Smyrna (Izmir), 39, 86
Sozopolis, 213
Spain, 94, 146, 148
Spatas, clan of, 188
Spata, Muriki, Albanian chieftain, 189
Sphrantzes, George, 7, 189
spoils of war (see also booty, plunder), 70, 223
Stefan Uroš II Milutin, Serbian ruler (1282–1321), 73, 164
Stefan Uroš III Dečanski, Serbian ruler, (1321–1331), 151
Stefan Uroš IV Dušan, Serbian ruler (1331–1355), 2, 39, 82, 110, 130, 131, 183, 184

Stefan Uroš V, Serbian ruler, (1355–1371), 2
Stenimachos (Asenovgrad), 171, 214, 216
stone-throwers (see also artillery weapons), 173, 174, 189
storehouses, 145, 146
Story of Belissarios, 66, 71, 72, 137 n. 2
Strategike ekthesis, 156
Strategikon of Maurice, 197, 206
strategoi of the themes, 83
Strategopoulos, Alexios, Byzantine general, 111
Strategopoulos, Constantine, 67
Strategopoulos, Michael, Byzantine general in Asia Minor, 24, 28
stratiotikon logothesion, 156
stratopedarches ton tzangratoron, 110
stratopedarches ton monokavallon, 87
Strymon river, 218
Sullivan, D., 181
sylloge tacticorum, 147
Synadenos, John, Byzantine general, 61, 141
Synadenos, Theodore, Byzantine general, 47, 60, 78, 102, 154
Synesius of Cyrene, 129
syntaxeis, 84
Syrgares, *lizios kavallarios*, 166
Syrgiannes, Byzantine general, 47, 48, 71, 81
Syrianos, Michael, head of a company of soldiers, 110

Tagaris, Manuel, Byzantine general in Asia Minor, 29, 47, 66, 69
tagma, 84, 85, 87, 91
Taitah, Tatar general, 165
Taktika of Leo VI, 63, 71, 147, 148, 197, 206–208
Tarchaneiotes, John, 78
Tarchaneiotes, Manuel, 69
Tarchaneiotes, Michael, 142
Tartaro, Arrigo, 32
Tatars, 104, 159, 164, 165, 169, 170
Teristasis, 161
Thasos, 189
Thebes, 103
thelematarioi, 95–96
Theodore I Laskaris, Nicaean emperor (1204–1221), 1, 17, 18, 20, 49, 77, 135, 201, 212
Theodore II Laskaris, Nicaean emperor (1254–1258), 1, 5, 6, 9, 10, 14, 18–20, 85, 105, 114, 115, 119, 120, 138, 141, 142, 144, 152, 154, 171, 172, 199, 209, 218
Theodore I Palaiologos, despot of Morea (1383–1407), 41–43, 65, 140
Theodore II Palaiologos, despot of Morea, (1407–1443), 97, 133
Theodore Doukas Angelos, despot of Epiros (1215–1230), 32 n. 77
Theodosius II, Byzantine emperor (405–450), 161
Theognostos, 14
Theoleptos, bishop of Philadelphia, 162
Thessalonica, 2, 14, 36, 37, 42, 43, 78, 88, 89, 91, 102, 103, 139, 145, 150, 154, 163, 167, 171, 173, 184, 190, 227
Thessaly, 1, 2, 20, 32, 35, 36, 72, 81, 105, 116, 140, 141, 167, 181, 217
Thomokastron, 181
Thrace, 34, 36, 37, 48, 77, 84, 92, 93, 104, 111, 137, 140, 142, 158, 159, 161, 164, 167, 168
Thrakesion, 84, 166
timar, 214
Tocco, family of, 103, 227
Tocco, Leonardo, 50
Toglu Turgan, Tatar general, 165
Tornikes, Constantine, 67
Tornikes, Michael, 48
Tountzas river, 165
Tourkopouloi, Turcopoles, 116, 213, 214, 217
tournament, 53–55, 57
Tralleis, 26, 178
Trebizond, empire of, 1
trebuchet, 172, 175, 186, 189, 192
trikavalloi, 89, 90
Trikokia, 178
Tripolis, 177
Turcoman mercenaries, 104, 211
Turcoman principalities, chiefdoms, 2, 4, 13, 14, 16, 24, 25, 27, 28, 30, 34, 38, 69, 79–81, 153, 162, 175, 203, 206, 207, 226, 227
Turkish allies of the Catalan Grand Company, 166, 168, 169
Turkish cavalry, 211, 214
Turkish mercenaries/allies, 102, 112, 116, 152, 153, 199–201, 203, 217, 218, 224
Turkish raids, 24, 34, 81, 87, 95, 161, 165, 176, 177, 214, 218

Turks, 8, 13, 16, 25–30, 38–40, 43, 44, 47, 48, 52, 56, 62, 64, 69, 70 n. 111, 77, 87, 97, 119, 123, 124, 126, 127, 137, 152, 162, 206–211, 214, 226, 227
typikon, 22
Tzakonike phylaxis, 167
Tzakones, 58, 108, 113, 114
Tzaousios, 169
Tzepaina, 67, 138, 142, 154, 216
Tzetzes, John, 172
Tzouloukones, 154
Tzouroulou, 154, 168
Tzykanion/tzykaton, 56

Umur, ruler of Aydın, 39, 40, 152, 214, 218
Urban militia, 128

Varangians, 113, 115
Vardariotai, 113–115
Varnakas, 188
Varvarenoi, company of soldiers, 109
Vatatzes, John, 33, 109, 161
Vatkounion, 144
Vatopedi monastery, 163
Velbužd, battle of (1330), 151, 152
Veles, 164
Venice, 106, 109, 110, 117, 131, 150
Vera (Pherai), 144, 164
Verilas, Basil, 90 n. 60
Veroe (Stara Zagora), 138, 172
Veroia, 102, 142, 167, 183, 184
Ververiakon, 171
Villehardouin, 214
Vizye (Vize), 89, 168, 169, 183
Vonitsa, 112, 188
Vrysis, 160

War of Troy, 51, 84
Western Europe, 3, 4, 37, 38–40, 51, 52, 60, 61, 149, 160, 187, 193, 219, 221, 223, 225
Western European authors, 63, 118
Western European knights/heavy cavalry, 54, 59, 65, 68, 195, 200, 201, 211–213, 224, 225
Western European mercenaries, 101, 117, 118, 131, 133, 135, 200
William II of Villehardouin, prince of Achaia (1246–1278), 22, 23, 52, 113, 201, 203
Winfield, D. 158, 192

Ximenez, Ferran, commander of the Catalan Grand Company, 169
Xyleas, Nicaean commander, 67

Yahsı Fakıh, 177
Yakub Alişir, ruler of Germiyan, 162
Yolande of Montferrat, 4

Zaccaria, Benedetto, 180
Zaccaria, Centurione, prince of Achaia (1402–1430), 98
Zaccaria Martino, 59, 161, 180
Zakythinos D., 98
Zographou monastery, 143

History of Warfare

History of Warfare *presents the latest research on all aspects of military history. Publications in the series will examine technology, strategy, logistics, and economic and social developments related to warfare in Europe, Asia, and the Middle East from ancient times until the early nineteenth century. The series will accept monographs, collections of essays, conference proceedings, and translation of military texts.*

1. Hoeven, M. van der (ed.). *Exercise of Arms.* Warfare in the Netherlands, 1568-1648. 1997. ISBN 90 04 10727 4
2. Raudzens, G. (ed.). *Technology, Disease and Colonial Conquests, Sixteenth to Eighteenth Centuries.* Essays Reappraising the Guns and Germs Theories. 2001. ISBN 90 04 11745 8
3. Lenihan P. (ed.). *Conquest and Resistance.* War in Seventeenth-Century Ireland. 2001. ISBN 90 04 11743 1
4. Nicholson, H. *Love, War and the Grail.* 2001. ISBN 90 04 12014 9
5. Birkenmeier, J.W. *The Development of the Komnenian Army: 1081-1180.* 2002. ISBN 90 04 11710 5
6. Murdoch, S. (ed.). *Scotland and the Thirty Years' War, 1618-1648.* 2001. ISBN 90 04 12086 6
7. Tuyll van Serooskerken, H.P. van. *The Netherlands and World War I.* Espionage, Diplomacy and Survival. 2001. ISBN 90 04 12243 5
8. DeVries, K. *A Cumulative Bibliography of Medieval Military History and Technology.* 2002. ISBN 90 04 12227 3
9. Cuneo, P. (ed.). *Artful Armies, Beautiful Battles.* Art and Warfare in Early Modern Europe. 2002. ISBN 90 04 11588 9
10. Kunzle, D. *From Criminal to Courtier.* The Soldier in Netherlandish Art 15501672. 2002. ISBN 90 04 12369 5
11. Trim, D.J.B. (ed.). *The Chivalric Ethos and the Development of Military Profes sionalism.* 2003. ISBN 90 04 12095 5
12. Willliams, A. *The Knight and the Blast Furnace.* A History of the Metallurgy of Armour in the Middle Ages & the Early Modern Period. 2003. ISBN 90 04 12498 5
13. Kagay, D.J. & L.J.A. Villalon (eds.). *Crusaders, Condottieri, and Cannon.* Medieval Warfare in Societies Around the Mediterranean. 2002. ISBN 90 04 12553 1
14. Lohr, E. & M. Poe (eds.). *The Military and Society in Russia: 1450-1917.* 2002. ISBN 90 04 12273 7
15. Murdoch, S. & A. Mackillop (eds.). *Fighting for Identity.* Scottish Military Experience c. 1550-1900. 2002. ISBN 90 04 12823 9
16. Hacker, B.C. *World Military History Bibliography.* Premodern and Nonwestern Military Institutions and Warfare. 2003. ISBN 90 04 12997 9
17. Mackillop, A. & S. Murdoch (eds.). *Military Governors and Imperial Frontiers c. 1600-1800.* A Study of Scotland and Empires. 2003. ISBN 90 04 12970 7
18. Satterfield, G. *Princes, Posts and Partisans.* The Army of Louis XIV and Partisan Warfare in the Netherlands (1673-1678). 2003. ISBN 90 04 13176 0
20. Macleod, J. & P. Purseigle (eds.). *Uncovered Fields.* Perspectives in First World War Studies. 2004. ISBN 90 04 13264 3
21. Worthington, D. *Scots in the Habsburg Service, 1618-1648.* 2004. ISBN 90 04 13575 8
22. Griffin, M. *Regulating Religion and Morality in the King's Armies, 1639-1646.* 2004. ISBN 90 04 13170 1
23. Sicking, L. *Neptune and the Netherlands.* State, Economy, and War at Sea in the Renaissance. 2004. ISBN 90 04 13850 1

24. Glozier, M. *Scottish Soldiers in France in the Reign of the Sun King.* Nursery for Men of Honour. 2004. ISBN 90 04 13865 X
25. Villalon, L.J.A. & D.J. Kagay (eds.). *The Hundred Years War.* A Wider Focus. 2005. ISBN 90 04 13969 9
26. DeVries, K. *A Cumulative Bibliography of Medieval Military History and Technology, Update 2004.* 2005. ISBN 90 04 14040 9
27. Hacker, B.C. *World Military History Annotated Bibliography.* Premodern and Nonwestern Military Institutions (Works Published before 1967). 2005. ISBN 90 04 14071 9
28. Walton, S.A. (ed.). *Instrumental in War.* Science, Research, and Instruments. Between Knowledge and the World. 2005. ISBN 90 04 14281 9
29. Steinberg, J.W., B.W. Menning, D. Schimmelpenninck van der Oye, D. Wolff & S. Yokote (eds.). *The Russo-Japanese War in Global Perspective.* World War Zero, Volume I. 2005. ISBN 90 04 14284 3
30. Purseigle, P. (ed.). *Warfare and Belligerence.* Perspectives in First World War Studies. 2005. ISBN 90 04 14352 1
31. Waldman, J. *Hafted Weapons in Medieval and Renaissance Europe.* The Evolution of European Staff Weapons between 1200 and 1650. 2005. ISBN 90 04 14409 9
32. Speelman, P.J. (ed.). *War, Society and Enlightenment.* The Works of General Lloyd. 2005. ISBN 90 04 14410 2
33. Wright, D.C. *From War to Diplomatic Parity in Eleventh-Century China.* Sung's Foreign Relations with Kitan Liao. 2005. ISBN 90 04 14456 0
34. Trim, D.J.B. & M.C. Fissel (eds.). *Amphibious Warfare 1000-1700.* Commerce, State Formation and European Expansion. 2006. ISBN 90 04 13244 9
35. Kennedy, H. (ed.). *Muslim Military Architecture in Greater Syria.* From the Coming of Islam to the Ottoman Period. 2006. ISBN 90 04 14713 6
36. Haldon, J.F. (ed.). *General Issues in the Study of Medieval Logistics.* Sources, Problems and Methodologies. 2006. ISBN 90 04 14769 1
37. Christie, N. & M. Yazigi (eds.). *Noble Ideals and Bloody Realities.* Warfare in the Middle Ages. 2006. ISBN 90 04 15024 2
38. Shaw, C. (ed.). *Italy and the European Powers.* The Impact of War, 1500– 1530. 2006. ISBN-13 978 90 04 15163 5, ISBN-10 90 04 15163 X
39. Biggs, D. *Three Armies in Britain.* The Irish Campaign of Richard II and the Usurpation of Henry IV, 1397-99. 2006. ISBN-13 978 90 04 15215 1, ISBN-10 90 04 15215 6
40. Wolff, D., Marks, S.G., Menning, B.W., Schimmelpenninck van der Oye, D., Steinberg, J.W. & S. Yokote (eds.). *The Russo-Japanese War in Global Perspective.* World War Zero, Volume II. 2007. ISBN-13 978 90 04 15416 2, ISBN-10 90 04 15416 7
41. Ostwald, J. *Vauban under Siege.* Engineering Efficiency and Martial Vigor in the War of the Spanish Succession. 2007. ISBN-13 978 90 04 15489 6, ISBN-10 90 04 15489 2
42. MCCullough, R.L. *Coercion, Conversion and Counterinsurgency in Louis XIV's France.* 2007. ISBN 978 90 04 15661 6
43. Røksund, A. *The Jeune École.* The Strategy of the Weak. 2007. ISBN 978 90 04 15723 1
44. Hosler, J.D. *Henry II.* A Medieval Soldier at War, 1147-1189. 2007. ISBN 978 90 04 15724 8
45. Hoyos, D. *Truceless War.* Carthage's Fight for Survival, 241 to 237 BC. 2007. ISBN 978 90 04 16076 7
46. DeVries, K. *A Cumulative Bibliography of Medieval Military History and Technology, Update 2003-2006.* 2008. ISBN 978 90 04 16445 1
47. France, J. (ed.). *Mercenaries and Paid Men.* The Mercenary Identity in the Middle Ages. 2008. ISBN 978 90 04 16447 5
48. Meyer, J. (ed.). *British Popular Culture and the First World War.* 2008. ISBN 978 90 04 16658 5
49. Jones, H., J. O'Brien & C. Schmidt-Supprian (eds.). *Untold War.* New Perspectives in First World War Studies. 2008. ISBN 978 90 04 16659 2

50. Burgtorf, J. *The Central Convent of Hospitallers and Templars.* History, Organization, and Personnel (1099/1120-1310). 2008. ISBN 978 90 04 16660 8
51. Villalon, A.L.J. & D.J. Kagay (eds.). *The Hundred Years War (Part II).* Different Vistas. 2008. ISBN 978 90 04 16821 3
52. González de León, F. *The Road to Rocroi.* Class, Culture and Command in the Spanish Army of Flanders, 1567-1659. 2009. ISBN 978 90 04 17082 7
53. Lawrence, D.R. *The Complete Soldier.* Military Books and Military Culture in Early Stuart England, 1603-1645. 2009. ISBN 978 90 04 17079 7
54. Beckett, I.F.W. (ed.). *1917: Beyond the Western Front.* 2009. ISBN 978 90 04 17139 8
55. Whetham, D. *Just Wars and Moral Victories.* Surprise, Deception and the Normative Framework of European War in the Later Middle Ages. 2009. ISBN 978 90 04 17153 4
56. Miller, S.M. *Soldiers and Settlers in Africa, 1850–1918.* 2009. ISBN 978 90 04 17751 2
57. Amersfoort, H. & P. Kamphuis (eds.). *May 1940.* The Battle for the Netherlands. 2010. ISBN 978 90 04 18438 1
58. Murdoch, S. *The Terror of the Seas*? Scottish Maritime Warfare, 1513-1713. 2010. ISBN 978 90 04 18568 5
59. Fagan, G.G. & M. Trundle (eds.). *New Perspectives on Ancient Warfare.* 2010. ISBN 978 90 04 18598 2
60. Klingelhofer, E. (ed.). *First Forts.* Essays on the Archaeology of Proto-Colonial Fortifications. 2010. ISBN 978 90 04 18754 2
61. Shean, J.F. *Soldiering for God.* Christianity and the Roman Army. 2010. ISBN 978 90 04 18731 3
62. Keene, J.D. & M.S. Neiberg (eds.). *Finding Common Ground.* New Directions in First World War Studies. 2011. ISBN 978 90 04 19182 2
63. Sharp, I. & M. Stibbe (eds.). *Aftermaths of War.* Women's Movements and Female Activists, 1918-1923. 2011. ISBN 978 90 04 19172 3
64. Corum, J.S. (ed.). *Rearming Germany.* 2011. ISBN 978 90 04 20317 4
65. Amersfoort, H. & W. Klinkert (eds.). *Small Powers in the Age of Total War, 1900-1940.* 2011. ISBN 978 90 04 20321 1
66. Potter, D. *Henry VIII and Francis I.* The Final Conflict, 1540–1547. 2011. ISBN 978 90 04 20431 7
67. Kyriakidis, S. *Warfare in Late Byzantium, 1204–1453.* 2011. ISBN 978 90 04 20666 3
68. Neiberg, M.S. (ed.). *Arms and the Man.* Military History Essays in Honor of Dennis Showalter. 2011. ISBN 978 90 04 20668 7

ISSN 1385–7827 brill.nl/hw

www.ingramcontent.com/pod-product-compliance
Lightning Source LLC
Chambersburg PA
CBHW071332230926
51849CB00002B/377
9789004206663